A Practical Guide to Human Research and Clinical Trials

Edited by

M U R Naidu
Professor and Head,

P Usha Rani
Additional Professor,
Department of Clinical Pharmacology and Therapeutics,
Nizam's Institute of Medical Sciences, Panjagutta,
Hyderabad.

CRC Press
Taylor & Francis Group
Boca Raton London New York

CRC Press is an imprint of the
Taylor & Francis Group, an **informa** business

PharmaMed Press
An imprint of Pharma Book Syndicate
A unit of BSP Books Pvt., Ltd.

First published 2013 by PharmaMed Press

Published 2019 by CRC Press
Taylor & Francis Group
6000 Broken Sound Parkway NW, Suite 300
Boca Raton, FL 33487-2742

ISBN 13: 978-1-4665-9117-2 (hbk)

Visit the Taylor & Francis Web site at
http://www.taylorandfrancis.com

and the CRC Press Web site at
http://www.crcpress.com

Distributed in India, Pakistan, Nepal, Myanmar (Burma), Bhutan, Bangladesh and Sri Lanka by **BSP Books Pvt., Ltd.**

Olivier BLIN
CIC-CPCET-Clinical Pharmacology Marseille
UMR 6193-CNRS -UNIVERSITE de la MEDITERRANEE : INCM
13385 Marseille Cedex 5, FRANCE
Tel : 33 (0)4 91 38 46 46 - Fax : 33 (0)4 91 47 21 40 – Mobile : 33 (0)6 24 306 802 – 91 (0)9951041000
mailto:olivier.blin@ap-hm.fr

Foreword

"Science", as a term remains a jargon unless and until its application reaches the masses; and if there is one discipline which has gained tremendous momentum in terms of having a direct impact to the society as a result of its allied research, it is Life-Science. Hence every basic research data generated are fast evolving into therapies in form of medications, vaccinations, stem cell transfusions etc. But this evolution has to rely on the clinical demonstration for the benefit of the patient, not only in terms of efficacy but also safety. Any brilliant scientific result, any innovative idea has to be evaluated in terms of these parameters before being spread for clinical use. Clinical trials ensure that relevant scientific results are useful for the Humans. Therefore, they require irreproachable data; and it's here that ethics comes into play.

Too many trials along with their respective investigators in a number of countries have been criticized for not respecting ethical rules and guidelines. This is not acceptable, irrespective of the countries, as clinical trials being a global business, has to be conducted with same level of quality and ethics throughout the world. Regulatory bodies such as the European Medicine Agency have done a tremendous work in collaboration with experts from the field to elaborate Good Clinical Practices, for publishing guidelines that applies not only in Europe but also in emerging countries. Ethics is an important part of this book and Prof. M.U.R Naidu has demonstrated a strong involvement in the field during his entire career as a clinical pharmacologist, drawing cartoons to explain the trials to volunteers largely ignorant about the study in order to make sure that they are aware of the pros and cons of the trials.

Conducting a successful clinical trial not only requires a strong basic knowledge, but also hands-on practical training. The book explains the intricate details of the subject to the reader by citing concrete cases, exercises and templates along with the theoretical aspects. With the co-authors, Prof. M.U.R Naidu has been able to address all important aspects of clinical trials - right from clinical research, drug development and quality to methodology, biostatistics and pharmacovigilance.

As one notices in the title, "Human Research" comes first, followed by "Clinical trials". This underlines the fact that clinical trials are of course essential but only when it methodologically approaches in finding answers concerning Huma research. Humans, in general, are not only the focal point of human research, bu with all its complexity and integrity, human research aims at benefitting the entir mankind.

I have known Prof. M.U.R Naidu for several years and knowing him fror such close quarters, I can ensure that this crucial dimension not only forms th chore of this book, but also forms an integral part in his own life.

Hence I feel privileged and take the pleasure in writing the forward of this boo and wish this will be a treasure for the inquisitive readers interested in research an its underlying ethics in general, and the patients and their families in particular.

Prof Olivier BLI

डा० एम० के० भान
Dr M. K. Bhan

सचिव
भारत सरकार
विज्ञान और प्रौद्योगिकी मंत्रालय
बायोटेक्नोलॉजी विभाग
ब्लाक-2, 7 वां तल, सी० जी० ओ० कम्पलेक्स
लोदी रोड, नई दिल्ली-110003

SECRETARY
GOVERNMENT OF INDIA
MINISTRY OF SCIENCE & TECHNOLOGY
DEPARTMENT OF BIOTECHNOLOGY
Block-2, (7th Floor) C.G.O. Complex
Lodi Road, New Delhi-110003

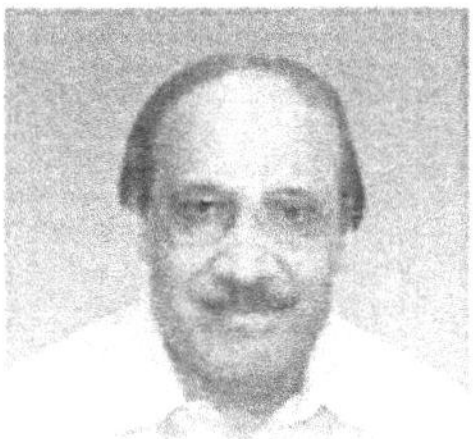

(M K Bhan)

Foreword

I am delighted that a new textbook entitled "A Practical Guide to Human Research and Clinical Trials" is being published. There is explosion of knowledge in Life Sciences which should increase the opportunities for developing new chemical and biological therepeutics. To capitalize on output of basic research, product development and evaluation for safety and efficacy, must happen in a manner consistent with the highest standard of ethics and quality. This requires a reliable, scientific, competent and transparent regulatory system and well trained, effective human resource for pre-clinical and clinical studies, backed by a world-class institutional framework. This book, written by highly experienced professionals is a welcome addition and will serve the purpose of producing effective well trained people. The book is comprehensive, covers drugs and devices, issues related to investigators, regulatory requirement and pre-clinical and clinical development. It has very interesting chapters on project design and planning of single and multiple centre clinical trails, good clinical research practice, ethics, quality assurance, data safety monitoring and archiving. Biomedical statistics has rightly received adequate attention. The book is characterized by clarity, simplicity and is easy to assimilate. It is designed to help acquire practical knowledge and skills. India has the opportunity to support development and health care evaluation of products within the counry and elsewhere in the globe. This is an enormous responsibility and an opportunity. The book is an excellent tool in the hands of those wanting to learn this art and science. I congratulate the authors for the excellent job done.

(M K Bhan)

Phone : +91 40 - 23489000
Direct No.: +91 40 - 23390933
Fax : +91 40 - 23310076
Email : director.nims@ap.nic.in

NIZAM'S INSTITUTE OF MEDICAL SCIENCES

(A University Established Under the State Act)
PANJAGUTTA, HYDERABAD - 500 082. A.P., INDIA

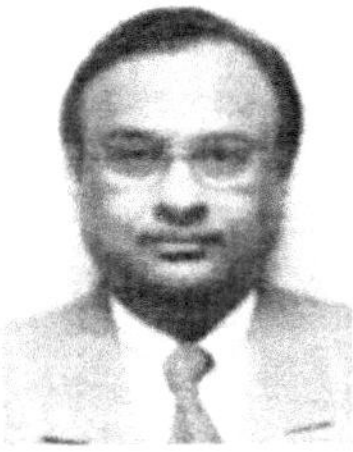

Prof. D. PRASADA RAO
M.S., M.Ch., (Cardiothoracic)
Director

Foreword

I am happy and privileged to know that Prof. M.U.R.Naidu and Dr.P.Usha Rani of Department of Clinical Pharmacology and Therapeutics, Nizam's Institute of Medical Sciences have authored a book (A Practical Guide to Human Research and Clinical Trials). I have gone through the book and found that it is extremely useful to the personnel involved in the Clinical Research. There is lot of scope for Clinical Research and Clinical trials in future. Since India is becoming a major medical hub, I strongly recommend this book for better care of our patients.

PROF. D PRASADA RAO

Preface

Many health professionals are expected to conduct their own trial, or to participate in trials by recruiting subjects. Every one who is conducting clinical research should have a sufficient understanding of the scientific and administrative aspect, including an awareness of the regulation and guidelines associated with clinical trials. Randomised controlled clinical trials have significant impact on the present practice of clinical medicine and other areas of health care. Subjects recruited in clinical research will range from healthy individual to patients from long standing illness. Ethical principles govern the choice and conduct of clinical research.

In the evaluation of therapies for existing and new diseases, clinical trials continue to be essential. This textbook addresses many issues of human clinical research and clinical trials, and provides comprehensive overview of the design, conduct and analysis of all the phases of clinical trials. It is aimed at health and other research professionals and students. The book can be used as teaching aid and as a reference guide. Important concepts are presented throughout the chapters with diagrams and check lists. I sincerely thanks all the authors who have contributed chapters for this edition

MUR Naidu

Contributors

Dr. M.U.R.Naidu,

Professor and Head,
Department of Clinical Pharmacology and Therapeutics,
Nizam's Institute of Medical Sciences, Panjagutta.
Hyderabad.

Dr. T.Ramesh Kumar Rao,

Professor,
Department of Clinical Pharmacology and Therapeutics,
Nizam's Institute of Medical Sciences, Panjagutta.
Hyderabad.

Dr. U.Shobha Jagdish Chandra

Additional Professor,
Department of Clinical Pharmacology and Therapeutics,
Nizam's Institute of Medical Sciences, Panjagutta.
Hyderabad.

Dr. P.Usha Rani,

Additional Professor,
Department of Clinical Pharmacology and Therapeutics,
Nizam's Institute of Medical Sciences, Panjagutta.
Hyderabad.

Dr. G.Venkat Ramana,

Scientist, Drug Development,
Natco Pharma Ltd.
Hyderabad.

Dr. Ahmed Abdul Mateen,

Research Associate,
Department of Clinical Pharmacology and Therapeutics,
Nizam's Institute of Medical Sciences, Panjagutta.
Hyderabad.

Dr. Y.Sridhar,

Research Associate,
Department of Clinical Pharmacology and Therapeutics,
Nizam's Institute of Medical Sciences, Panjagutta.
Hyderabad.

Ms P.V.N.Sridevi

Project Lead,
Sristek Consulting Pvt Ltd.
DLF Cyber City, Block-3, 8th Floor, Gachibowli
Hyderabad.

Contents

Chapter 4

Phases of Clinical Trials

Chapter 5

Phase 0 - Microdosing Studies

Chapter 6

Clinical Research Planning

Chapter 7

Clinical Research Design

Chapter 8

Clinical Research Protocol

Chapter 9

Designing Case Report Forms

hapter 13

Good Clinical Research Practice

hapter 14

Role and Responsibilities of Sponsor

hapter 15

Role and Responsibilities of Principal Investigator

Chapter 16

Ethical Consideration in Clinical Research – Special Reference to Developing Countries

Chapter 17

Informed Consent in Clinical Research

Chapter 18

Monitoring in Clinical Trials

Chapter 19

Quality Assurance in Clinical Research

Chapter 20

Pharmacovigilance and Drug Safety

Chapter 21

Data and Safety Monitoring Board and Monitoring Plan

Chapter 22

Standard Operating Procedures

Chapter 23

Archiving Clinical Research Documents

Chapter 24

Evidence Based Medicine

Chapter 25

Clinical Research Data Management

Chapter 26

Clinical Biostatistics

(a) Biostatistics for Clinical Researcher

CHAPTER 1

Clinical Research - Clinical Investigator's Perspective

M.U.R. Naidu and P. Usha Rani

.1 Introduction

:linical research / trials require a substantial commitment of both time and effort s participation may continue for months or years. Though financial rewards re limited for participating in a clinical trial, there are other incentives like:

- The chance to collaborate with other clinical investigators.
- Experience of clinical research methods for study site personnel.
- Opportunities to enhance knowledge of the disease and the treatment under investigation.
- In some trials, exposure to new investigative techniques or access to specialized equipment or facilities.

The scientific, practical and financial implications should be weighed before onsidering participation in a clinical research / trial.

It may be prudent to consider and address the following aspects before an ıvestigator decides to undertake a clinical research project.

1.2 Scientific Aspects and Practical Aspects

- It is important to determine that the trial addresses a question of clinic and scientific relevance. Some trials permit a commercial sponsor to app for an extension to the marketing authorization of an investigational produ but do little to improve the understanding of disease and therapeutics. T same question is sometimes addressed by different sponsors and tl proposals may be redundant as the information is readily available.
- Investigator must discuss clinical trial design and statistical issues a should assess whether the protocol is likely to confirm or refute a trial hypothesis. An important consideration for an investigator is whether tl required number of clinical research participants stated in the protoc can be enrolled in the defined time frame. The requirements for follow of patients must also be considered, e.g., the duration and frequency follow up and the data to be collected.
- Investigator must carefully assess the eligibility criteria as the number patients at a study site, who might be eligible for a trial may l overestimated. Some study sites keep records of admission to tackle th problem. More experienced study sites can often use information fro previous trials that had similar eligibility criteria as a guide. Investigato should be confident that sufficient eligible patients will be recruited f the study. It is advisable to survey and maintain a log of potentially eligib study participants. This will provide a guide to estimate recruitment rate
- An investigator's primary obligation is to protect the welfare of his/h participants. An investigator should consider how much time he/she wi have to devote to the study, and compensation they may receive. Will tl patients be required to undergo investigations that would not be considere part of routine care, and if they have to, will they be painful or put tl patients to additional risk? The Ethics committees pay particular attentic to these issues when reviewing a research study protocol.
- Data collection is a time consuming task, especially if case report form ask for vast amount of data, much of which may contribute little to tl study results. It is important to address how the data will be amassed ar recorded. Procedures for processing adverse events may be very tin consuming and this may not be clear at the outset of the trial.
- It is essential to ensure that the study site and institution have staff availab to run the trial and that these people are sufficiently experienced in clinic trials before committing to participation in a clinical trial. This is particularl true of the study coordinator and any co-investigators who will be assigne the task of running the trial on a day to day basis. Curricula vitae is usual requested by the study sponsor because it is a requirement of t International Conference on Harmonization, Good Clinical Practice (ICI

GCP) and Indian GCP guideline, to ensure staff are appropriately qualified and adequately trained. It is important to assess the amount of time required for patient enrolment, data collection and follow up and to consider whether study staff will have time to fulfil these obligations.

When calculating the staff costs, it is important to consider the following questions:

- How many personnel will be needed to fulfil the requirements of the protocol?
- Are the salary levels of team members identified?
- What proportion of their time will be spent on the study?

- It is further essential to address the issue of post-trial access of study medication for the participating subjects.
- It is important that investigators sort out publication related issues before start of any study and the same be clearly mentioned in the clinical study agreement.

.3 Financial Considerations

ayments to study sites vary from study to study but the principles applied when aching a decision to participate in a trial should remain the same.

- Most institutions now may set up a specific department, or appoint an external group to review studies and ensure that local health service resources are not used inappropriately to subsidize research. Prior to submitting a study for review by such groups, investigators should perform a financial review themselves and ideally develop a local budget. Local costs are usually predictable. However, it is important to remember that there may be many 'hidden' costs of drug storage, ethics committee review and provision of laboratory certificates. If specimens are to be sent to a central laboratory it is important to ensure that costs of local processing, storage and shipping are covered.
- Investigators should also consider the structure of the payment schedule. Most research studies pay a set amount of money per subject recruited, but the timings of such payments may vary. Possible payment schedules are:
  - One payment per patient when the patient is recruited.
  - One payment per patient once certain pre specified data have been declared clean i.e. from omissions or inaccuracies, and hence appropriate for statistical analysis.
  - One payment per patient at the time that the entire dataset has been received and declared 'clean'.

- A structured payment schedule whereby set amounts are paid at recruitment for the receipt of data from various follow-up visits and at the time that the entire dataset has been declared 'clean'.

The investigator should make sure that all eventualities have been considered when agreeing to undertake a study where the payments are based upon receipt of 'clean' data and/or are tied to follow up visits. Investigators must be aware that there is a potential for payments to be withheld if there is poor or no recruitment. Further, if payments are dependent on patient follow-up, then it is vital that these follow-ups take place. Also, contracts should be examined to ascertain what will happen in circumstances that are beyond the control of the investigator e.g., death of a patient.

Many protocols are amended one way or the other during the course of the study and extension to recruitment periods is a real possibility. Investigators should be realistic in the financial planning of a project whilst ensuring that the costs are adequately covered by the contract.

Decide! To participate or not

Finally, clinical investigators may choose to take part in a study because they appreciate the scientific or clinical importance of the research project. However, they may be willing to subsidize research that is not adequately funded, though this is less usual. Equally possible is the fact that they may decide to participate simply because they feel that the study is well funded.

Suggested Reading

1. Spilker B. Guide to clinical trials. New York 1991, Raven Press.
2. Tony Craig. Factories, call centres —— and now clinical trials outsourced to India. National Review of Medicine, July 30, 2004; 1 (4).

CHAPTER 2

Medical Device Development, Process and Regulation

M.U.R. Naidu and P. Usha Rani

2.1 Introduction

Recently, with rapid growth in scientific and engineering technologies, large number of medical devices are developed. Both basic, biomedical and bioengineering research with advances in the physical sciences and mathematics significantly contributes to knowledge of medical device development. Compared to drugs, devices are more heterogeneous in terms of design, use and purpose. There are roughly 2000 different types of medical devices and more than 50,000 products. Usually based on engineering problem solving by individuals, medical devices are invented. After a medical device product is invented, a patent application is filed. As compared to patent protection to the pharmaceuticals, value of patent protection in medical device is less evident. To invent around a patent, device development time is generally shorter and easier.

Competition from other manufacturers is very less in case of medical devices, due to large capital cost and large scale investments. Clinical researchers, physician and surgeons who are the potential users play an important role in device development. There is some variation in clinical evaluation with medical devices as they are more heterogeneous.

2.2 Categories of Medical Devices

Medical devices range from low risk devices to moderate and high risk devices. The development and manufacture of medical devices under a controlled environment is essential to assure that the device will be safe and effective in accordance with its intended use.

Medical devices are universally classified mainly into three categories:

Class I which includes instruments such as tongue depressors, those that do not support or sustain human life and not having unreasonable risk of illness or injury. Class I type devices constitute to about 30% of all the medical devices.

Class II about 60% of devices are in Class II, and involve some degree of risk and are subject to regulatory defined performance standards such as X-ray devices.

Class III all medical devices that are life supporting or sustaining, and are useful in preventing the impairment of health or which have potential risk of injuries or illness, are classified as Class III medical devices. Before the regulator grants the marketing approval, manufacturer has to demonstrate safety and efficacy of class III medical devices. Approximately 10% of medical devices are in this class and include artificial heart, DNA probes or laser angioplasty devices.

In India, in June 2009, the Drug Consultative Committee (DCC) and the Drug Technical Advisory Board (DTAB) approved the new guidelines for India's medical device development and regulation formally. The same is posted on Central Drug Standard Organization (CDSCO) website under Schedule M-III section. Thus Schedule M-III results in regulating medical devices as a separate entity from drugs.

According to Schedule M-III, medical devices are divided into four classes – A, B, C and D according to their risk level. Class A will include low risk devices such as thermometers. Low to moderate risk devices like hypodermic needles will fall under Class B. The next Class C will cover moderate to high risk devices like ventilator and High risk devices, heart valves and pace makers will makeup Class D.

2.3 Evaluation of Medical Devices

In comparison with drugs, Institutional ethics committee (IEC) plays a more important role in device evaluation. The IEC determines whether a device possesses a significant risk on the basis of an investigational plan. Unlike pharmaceuticals, the final version of a medical device is often not created denovo, instead a device prototype is usually modified as a result of initial evaluation.

efore the device is used properly and efficiently long period of learning is cessary than drugs. Clinical testing of devices usually first involves a pilot ge in which prototype design and material are tested further. In addition to nical evaluation this stage also involves technical evaluation eg; testing of ctrical and mechanical components of device. Evaluation of the final initial oduct can be initiated after the technical development has been completed d product is stabilized. Risks and benefits assessment of device may not reflect early phase evaluation, decision for further development are sometimes difficult clinical evaluation usually reflect risk and benefit at a fixed time point.

For medical device evaluation, a distinction needs to be made between agnostic and treatment devices. There is no direct patient benefit from diagnostic vice, but may benefit in terms of clinical utility. These devices must be evaluated r diagnostic accuracy, diagnostic and therapeutic impact, and patient outcome. ne main measures of diagnostic performance are sensitivity and specificity. In inical practice, additional measures like predictive value of a positive test and edictive value of negative test play an important role. Traditionally most devices aluation lack randomized clinical trials, which may be due to inherent aracteristics of device development that makes the classical randomized ntrolled trials more difficult to conduct. Both patient and investigator blinding ay also cause more difficulties with device trial. However, upto some extent as can be eliminated by method in which one investigator may insert or implant e device, while other will evaluate its benefit and risks. Generally, the sample ze of device trials are much less compared to drug trials. During the clinical aluation of medical devices, much industrial effort is directed towards scaling the production. In general, device manufacturing is governed by GMP gulations and both national and international standards may also exert important fluence. Based on the out come of clinical investigations, a device may be proved for marketing. Involvement of practicing clinician is required from the rliest stage of device development.

.4 Regulatory Process

nowledge and compliance with regulatory requirements is a key to success in velopment and marketing of medical devices

(i) Regulatory approval process in US – Medical devices are regulated in the US by the Center for Devices and Radiological Health (CDRH of FDA). The FDA/CDRH mandate is to promote and protect the public health by making safe and effective medical devices available in a timely manner. The standard of demonstrating safety and effectiveness is determined in part by the risk associated with the device in question.

Most class I devices before there marketing do not require formal FDA review. Before the class II device may be marketed, it requires FDA clearance of a premarket notification application (PMA). In th application, the medical device manufacturer must provide data demonstrate that the new device is substantially equivalent to a legal marketed device. Most of Class III devices require FDA approval of PMA before they can be legally marketed. It generally requires clinic data demonstrating reasonable assurance that the device is safe a effective in large population. Clinicians are frequently involved in creati of the device concept and are active members in the team conducti early animal studies.

(ii) Regulatory Approval process in European Union – The European Uni System relies on notified bodies which are the commercial organizatio to regulate the medical devices in European countries. Notified bodi issues the CE mark which is the official marking required for certa medical devices. Similar to FDA / CDRH notified bodies perform ta of certification, device type designation, assessment, verification of quali systems and review of design dossier for high risk devices.

(iii) In India, the Central Licensing Approval Authority (CLAA), branch CDSCO, will serve as the main regulatory body for medical device The CLAA will classify medical devices. Hence all medical devic need to undergo conformity assessment to assure quality and standa before they are marketed in India. In order to assure the quality ar specifications of medical devices are configured appropriately the CLA will confer to the regulatory standards as laid by the Bureau of India Statistics (BIS) and International Organization for Standardization (ISC Medical devices sold in this country should bear a ICAC mark (India Conformity Assessment Certificate) to indicate their conformity wi provisions of this schedule to enable their marketing.

Conformity assessment procedures differ between the class of device. Cla A device manufacturer needs to register with the CLAA, and conformit assessment can be performed by manufacturers. However, for Class B, C ar D devices, the CLAA, in consultation with the BIS will publish a list of notifie bodies authorized to perform conformity assessment. Further, the class D devic manufacturing facility may also be inspected jointly by CLAA and State Licensir authority. In case of Class C and D devices, additional information and clinic investigation may be required.

Whenever possible, sampling of medical devices shall be carried i accordance of the procedure laid down in Drugs and Cosmetic Act 1940. I case the nature of the devices is such that above procedure cannot be adopte CLAA may take sample for evaluation from experts in the field or take an other measure to verify the claim of the manufacturer.

Suggested Reading

1. Maxwell RA. The state of the art of the science of drug discovery – an opinion. Drug Development Research 1984;4:380.
2. Keit Summerhayes. The challenges of conducting medical device studies – ICR publication.
3. Zbinden G. Current trends in safety testing and toxicological research. Natur-wisenschaften 1982:69:255-259.
4. Kaitin Kl, Richard BW, and Lasagne L. Trends in drug development: The 1985-86 new drug approvals. Journal of Clinical Pharmacology 1987:27:542-548.
5. Lasagne L and Bearn AG. Innovation and Acceleration in Clinical Drug Development. Roundtable 2:Early clinical trials – Phase I-IIA: Optimal designs for maximum information. New York: Raven Press, 1987,pp 95-103.
6. Aaron V Kaplan, Donald S Baim, John J Smith et al. Medical Device Development – From Prototype to Regulatory Approval. Circulation 2004: 109: 3068-72.
7. www.cdsco.com
8. www.reuters.com

CHAPTER 3

Pre-Clinical Drug Development

Abdul Mateen, Y. Sridhar, P. Usha Rani and M.U.R. Naidu

3.1 Introduction

Finding a novel drug is a complex process, whether the new drug is a small organic molecule or a macromolecule such as protein or oligonucleotide it requires scientific expertise of a number of disciplines and an enormous amount of time, money for its development. Historically, the main source of biologically active compounds used in drug discovery programs has been the natural products isolated from plants, animals or fermentation sources, but with the advent of newer technologies, the scenario has significantly changed, thereby, not only are more number of chemically synthesized compounds screened but, the time required for generation of appropriate leads has been shortened to a great extent. However, it is seen that for every 100 compounds screened for pharmacological activity in animal models, only one has the necessary biological activity and safety for evaluation in humans. Of those compounds tested in humans, only 1 in 40 is successfully brought to the market place. The reason for this poor rate of success may be, insufficient time and resources that are put into characterizing a lead candidate for its pharmacological activity, pharmacokinetic properties and safety profile in animal models to select the compound with desired attributes for development. Thus, a through screening of the lead compound for its biological activity, metabolism and toxicity will help to eliminate poor candidates from the scovery pipeline much earlier, saving time and money. This chapter will describe e basic aspects involved in preclinical drug development, with a greater emphasis n preclinical safety assessments.

.2 Overview of Drug Discovery and Role of basic Sciences

he development of any potential drug begins with years of scientific study to etermine the biochemistry behind a medical problem for which the medical tervention is sought. The result of these would be determination of specific ceptor targets that, may be modulated to alter their activity. The identification f these therapeutic targets requires knowledge of disease etiology and the iological system associated with it. Molecular biology has revolutionized the rocess of drug discovery. Earlier scientists searched for new targets employing long and costly process of trial and error. Today collective contribution of enomics, proteomics and bioinformatics allows for much more rapid and precise iscovery of those genes and/or proteins involved in the etiology of certain iseases.

The modern drug discovery pipeline is outlined in Fig. 3.1. Once the therapeutic rget is identified through one of the areas of basic sciences the next step ould be the generation of one or more lead compounds that, would interact ith the therapeutic target so as to produce the desired therapeutic effect. In rder to discover compounds whose pharmacological properties are likely to ave the required therapeutic effects large numbers of compounds are extensively creened and a lead compound is identified. After the lead is selected it is haracterized, synthesized and taken up for further development. This evelopment process starts with evaluating the efficacy and selectivity of the st compound in suitable animal models, simultaneously the drug developability udies as well as preliminary toxicological and absorption, distribution, metabolism, nd excretion (ADME) studies are also performed.

Once the results of preliminary studies are obtained the potential drug candidate s subjected to detailed preclinical studies, which mainly involves detailed hysicochemical studies, ADME studies, safety pharmacology and toxicological tudies. These aspects are dealt in detail in the following sections.

.3 New Approaches in Drug Discovery

ombinatorial Chemistry

ombinatorial chemistry is one of the new methodologies developed by esearchers in pharmaceutical industry to reduce the time and cost associated ith developing effective and competitive new drugs.

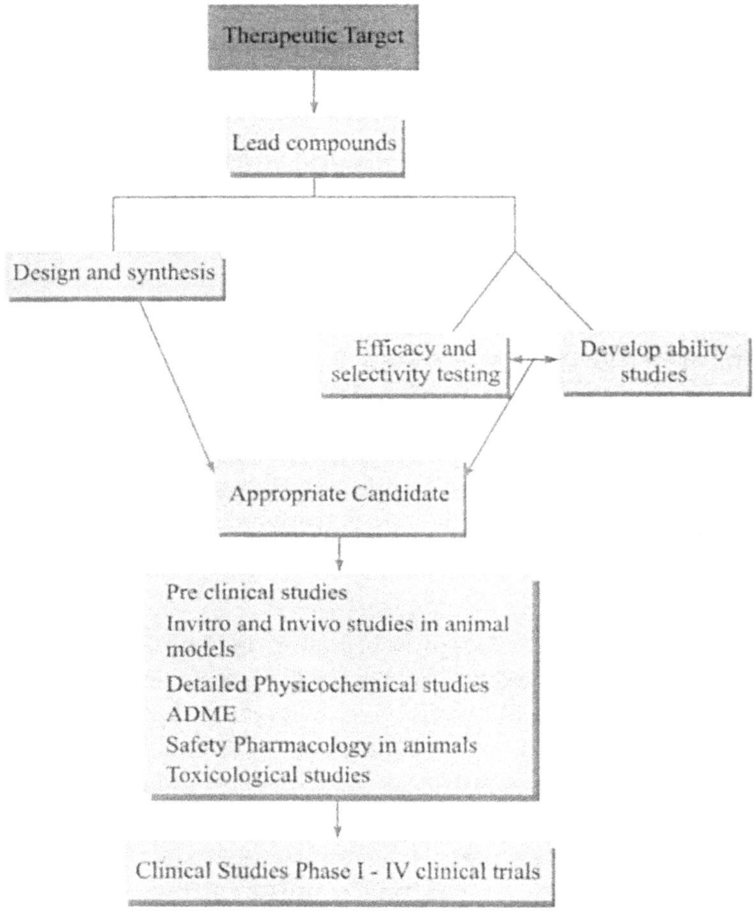

Fig. 3.1 Out line of drug development process.

Combinatorial chemistry (or combichem) is an innovative method o synthesizing many different chemical substances at the same time. It contrast with the time consuming and labor intensive methods of traditional chemistry where compounds are synthesized individually, one at a time. Combichem i used to synthesize large number of chemical compounds by combining sets o building blocks. Each newly synthesized compound composition is slightly differen from the previous one. A traditional chemist can synthesize 100-200 compound per year, but a combinatorial robotic system can produce thousands or million of compounds that can be tested for potential drug candidates in a high throughpu screening process. Over the last few years, the combinatorial chemistry ha emerged as an important and exciting new paradigm for the drug discovery.

ombinatorial Chemistry Techniques Include

1. Initiating a large number of similar reactions that are taking place simultaneously in multiple reaction vessels, where mixing of reagents and post reaction work is automated.
2. Combining a mixture of reactants in the same reaction vessel to produce a library of products in a single vial and testing overlapping libraries to find those mixtures that contain active components
3. Using solid phase chemistry, with the reactant attached to a support, and multi step synthesis, in which appropriate reagents are passed over the reactant on the support to prepare single or multiple products that are cleaved from the support and split and mixed so that subsequent steps have multiple starting material.

With these techniques, libraries have been created for many classes of New emical Entities (NCEs) some of which include â-lactams, benzodiazepines, iazolidinediones, prostaglandins, imidazole, peptides, oxazoles, alkaloids and parins.

igh Throughput Screening

igh throughput screening, often abbreviated as HTS is a scientific perimentation, especially used in drug discovery. HTS is a relatively recent novation, made feasible through modern advances in robotics and high speed mputer technology, with a combination of modern robotics, data processing d control software, liquid handling devices and sensitive detectors, HTS allows researcher to effectively conduct millions of biochemical, genetic, or armacological tests in a short period of time. Through this process one can pidly identify active compounds, antibodies or genes, which modulate a particular omolecular pathway. The results of these experiments provide a starting point r drug design and for understanding the interaction or role of a particular ochemical process in biology.

High throughput screening uses a computer generated approach to collect a rge amount of experimental data, usually observations about how some biological tities react to exposure to various chemical compounds in a relatively short ne. A key piece of HTS equipment is a plate, a small container usually made of astic that features a grid of small, open divots called wells. Most of the wells ntain experimentally useful matter, often a solution of dimethyl sulfoxide)MSO) and the other chemical compound that is different for each well across e plate.

First step of HTS assay involves filling of each well of the plate with some ological entity for which the experimentation has to be carried out, usually ese are proteins, some cells, or an animal embryo. After incubation time which allows the biological matter to absorb, bind or react with the compound in ea well, measurements are taken across all the plate's wells either manually or a machine, usually automated machines are employed to read the wells.

Depending upon the results of this first assay, the researcher can perfor follow up assays by "cherry picking" liquid from the wells that gave interesti results (known as "hits") into new assay plates, and then re-running t experiment to collect further data on this narrowed set, each time confirmi and refining the observations.

Automation is an important element in HTS usefulness. A specialized robot often responsible for much of the process over the lifetime of a single ass plate, from creation through final analysis. A HTS robot can usually prepare a analyze many plates simultaneously, further speeding the data-collection proce HTS robots currently can test up to 100,000 compounds per day.

3.4 Pharmacological Screening

In-vitro Screening

In-vitro Screening methods are robust and cost-effective, and have fewer ethic implications than whole entire-animal experiments. In-vitro assays monitor surrogate readout. Examples for such a read out are the catalytic action of isolated enzyme, the binding of an antibody to a defined antigen, or the growth an engineered cell line. On the other hand *In-vitro* Cell-based assays range their complexity from simple cytotoxicity assays or cell growth to reporter ge assays that monitor activation or up regulation of certain genes or their ge products.

In-vivo/Animal Models

In-vivo testing involves whole organisms. It assesses both pharmacology ar biological efficacy in parallelsimultaneously. Animal models have specif characteristics that mimic human diseases. The technologies for the creation transgenic animals, where certain genes are either deleted, modulated, or adde have progressed tremendously in the last decade. As a consequence, the predicti power of animal models for human disease and pharmacology is improvin Human biology and disease is so complex that for many diseases pharmacological parameters, the human remains the definitive model. For son disease, e.g., hepatitis C, adequate animal models still do not exist.

Drug Developability Studies

Once the lead is generated, characterized and the initial studies indicate that t compound or the class of compounds is havinghas the potential therapeut benefits, additional drug Developability studies are carried out before definiti drug development program is begun. These Develop ability studies may unrav rtain limitations that might have to be resolved before the commencement of eclinical and clinical studies. Fig. 3.2 gives and overview of various disciplines drug Develop ability studies.

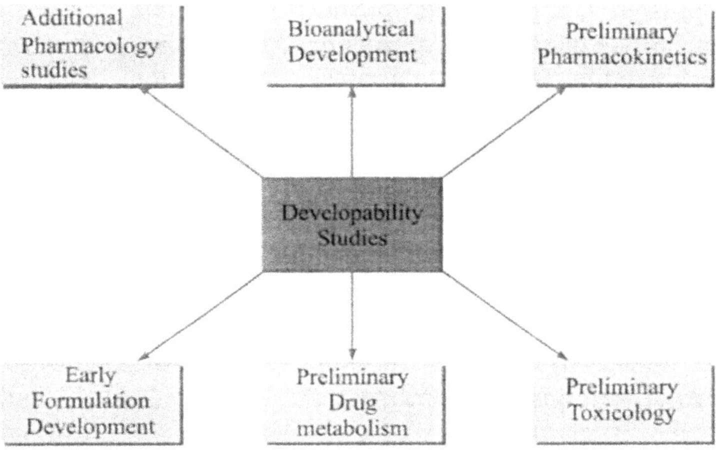

Fig. 3.2 Overview of Develop ability studies.

1. Additional Pharmacology Studies

In vitro and *In vivo* models in preliminary pharmacology evaluation might have showed that the lead compound has potential therapeutic benefit, but additional pharmacology studies may be needed to further characterize the dose, response curve using proposed clinical route of administration, frequency of administration and biological fluid concentration. Wherever possible these studies are conducted in at least two animal species to showstudy the biological response. ED_{50} dose and therapeutic ratio is also calculated. If the therapeutic ratio is one or less, then the compound is most likely to elicit adverse events in addition to the beneficial response. On the other hand a therapeutic ratio of five or more indicates that the compound is most likely to produce the pharmacological response before causing dose-limiting toxicity. Unless the compound under test is to be used as a single dose therapeutic, these additional pharmacological studies are conducted with steady state dosing. These additional pharmacological studies will eventually helps in selection of dose levels, route and frequency of administration for preliminary and definitive toxicology studies and also for initial phase I human safety and tolerance studies.

2. Bioanalytical Method Development

A Bioanalytical method is defined and characterized for quantification of the compound in biological fluid. For the compounds with a molecular weight of less than 2000, instrumental methods such as HPLC, GC or

LC-MS with a wide variety of detectors ultraviolet, fluorescence, flam ionization, electron capture and mass spectrometry are generally use For macromolecules like peptides, proteins or oligonucleotides ELIS (Enzyme linked immunosorbent assay) or RIA (Radio immuno assay) a employed. The developed bioanalytical method is characterized demonstrate the range of reliable results, the lower and upper limits quantification, specificity, accuracy and precision. Using appropria techniques samples are analyzed in biological fluids over a wid concentration range, and the potential interference with the matrix (bloo plasma, serum, urine) is also detected. Stability studies are also conducte to ensure that the compound does not degrade in blood during processin to obtain plasma or serum and under storage conditions (room temperatur refrigerated, frozen at -20^0 C or -80^0 C).

3. **Preliminary Pharmacokinetics**

The animals species used for preliminary pharmacokinetics study a generally the same as used in pharmacological evaluation, most likely th rodents. These preliminary pharmacokinetics studies help confirm th the bioanalytical method developed is useful in characterizing th absorption and disposition profiles of the compound. An example of suc pharmacokinetic study may entail dosing of rats with IV bolus injection a dose level between 25% and 50% of the pharmacologically active dos and dosing rats per oral (P.O) at the pharmacologically active dose. Bloo samples are collected at definite time points from each rat, processed t obtain the desired biological fluid and analyzed. The plasma concentratio versus time profiles after IV dosing give preliminary information on th distribution and disposition kinetics of the compound, where as the P. plasma concentrations versus time profile provides information on th absorption kinetics and absolute bioavailability.

4. **Early Formulation Development**

The early formulation development studies are necessary to define a acceptable formulation and are dependent on the clinical route c administration. For example a compound that has to be administered b IV injection, its formulation needs to be such that the compound does nc precipitate when administered and has minimal local toxicity. Compound that are highly lipophilic and havinge limited aqueous solubility are mo likely to have these types of problems. A low extent and high variability i absorption will be a problem for compounds that are to be administere by oral, subcutaneous or dermal routes. For cCompounds that hav absorption of less than 25% of the administered dose or a variability absorption of more than 100% of the amount absorbed, requir development of formulations with absorption enhancers or solubilizers t improve the drug delivery profile.

5. Preliminary Drug Metabolism Studies

The drug metabolism studies are conducted to determine the distribution, disposition, metabolism and elimination of the compound. These studies are generally conducted with a radiolabelled compound. These studies can be conducted with a less than desirable radiolabel such as ^{125}I on a protein or ^{3}H at a potentially exchangeable site on new chemical entity.

6. Preliminary Toxicological Studies

Before the definitive toxicological studies are initiated preliminary toxicological studies are conducted to determine the safety profile of the compound as well as finding out the no toxic effect dose, MTD, potential organs of toxicity and potential biomarkers to detect and track toxic events. These studies are generally conducted in same species as used in pharmacological evaluations. The lowest dose that has no toxicity or an acceptable level of toxicity is compared with the dose that gives the desired pharmacological response in the same animal species and the therapeutic ratio or index is obtained for that species. These preliminary toxicological studies need not be conducted as per the GLP requirements, but are generally designed and conducted as close as possible to the tests conducted for definitive GLP toxicological studies. Few examples of preliminary toxicological studies that are generally conducted are:

(i) *Acute Single-dose Tolerability Studies*

These studies are generally conducted in two relevant mammalian species, a rodent and a non-rodent using equal number of males and females. A single dose at various dose levels of the compound is administered by the proposed clinical route of administration to evaluate the qualitative and quantitative single dose toxicity. The animals are observed for 14 days after dosing and all signs of toxicity with time of onset, duration of symptoms and reversibility are recorded. The highest no toxic effect dose and the highest nonlethal dose are also determined

(ii) *Dose Range Finding Studies*

The primary aim of these studies is to determine the MTD. The doses for definitive toxicology studies are also defined in dose range finding studies. The dose ranges for these studies usually consist of four dose levels, the highest level being the dose that did not cause acute toxicity. These studies are performed in each species that are proposed for definitive toxicological studies. For rodents the number of animals usually will be 6 to 10 per dose group (3 to 5 animals per sex). For non rodents normally beagle dogs or nonhuman primates are used, commonly 4 animals (2 per gender) per dose. Endpoints for dose range finding studies generally include but are not limited to, weight loss, activity changes, clinical chemistry changes, histology and pathology evaluations.

3.5 Pre-Clinical Safety Assessment

Before the start of human trials, a drug candidate is subjected to a number c preclinical studies to further define and characterize its safety profile. Pre clinica safety testing is one of the important steps in the development of any drug. Th requirements of such testing are the subject of government regulations and ma vary in details, but not in general principles, in different countries around th world. The approach to preclinical safety testing of a drug is multidisciplinar and requires the expertise of members from different fields of science. All th preclinical studies have to be conducted according to GLP regulations. Howeve it is recognized that in some studies (studies employing specialized test system for example in case of biopharmaceuticals) it may not be possible to fully compl with the GLP, in such cases a proper justification has to be provided and th deviations hasve to be documented.

Animals used in Preclinical Safety Assessments

Different species of animals are used in preclinical safety assessments. Th choice of animal species depends on the test that has to be performed. The mos common rodents generally used are rat and mouse, where as beagle dog is th most common non rodent used in preclinical safety testing. However, other anima species like hamster, ferret, guinea pig, rabbit, primates (monkeys, marmose squirrel monkey, baboon) micro pig can also be used. The choice of anima species depends on the test that has to be performed, for example a non roder like dog is preferred for chronic toxicity test, but in case of reproductive toxicit assessments rabbit is preferred.

3.6 Types of Toxicological Studies

Before an application for phase I study is filed with the regulatory authoritie various toxicity studies have to be conducted in animals. There are differen types of toxicity studies that have to be conducted for determining the safety o the potential drug candidate. They are:

1. **Sub-Chronic and Chronic Toxicity Studies:** Subchronic toxicity fo the potential drug candidate has to be evaluated in two species, one roder and one non rodent. If the potential drug candidate is to be administered i humans for a period of more than 3 months, chronic toxicity studies ar essential and have to be evaluated in two species. The two basic reason for conducting chronic toxicity studies are to produce a toxic effect and t define a safety factor. These studies provide a dose response relationshi to effects resulting from prolonged exposure to the drug candidate an should reveal adverse effects that require a long exposure to be expresse

   The two most commonly used species in sub chronic and chronic toxicolog studies are rat and dog. The most common rat that is used in pharmaceutica and biotechnological industry is the Charles River CD rat, where as beagle dog (pure bred or specifically bred for research) is the common non rodent used. Rabbit can also be used as the non rodent species for sub chronic toxicity study. For rodent subchronic toxicity studies, usually 10-25 animals/sex/group are used. Smaller groups are generally used for shorter studies (2-4 weeks), whereas larger groups for long term studies (more than 13 weeks). In case of non rodent generally 3-6 animals/sex/group are used, depending on the length of the study and expected toxicity profile.

The duration of subchronic and chronic toxicity studies is related to duration of proposed clinical trials and intended use in humans. ICH guidelines suggests the minimum duration of toxicity studies needed to support phase I,II and III clinical trials in which humans are to be exposed to the drug candidate for varying duration as, shown in table below;

Minimum duration of Toxicity study

Duration of Clinical Use	**Rodents**	**Non-rodents**
Single dose	2-4 weeks	2 weeks
Up to 2 weeks	2-4 weeks	2 weeks
Up to 1 month	1 month	1 month
Up to 3 months	3 months	3 months
Up to 6 months	6 months	6 months
GreaterMore than 6 months	6 months	Chronic

Generally a six month chronic study in a rodent and a non-rodent is sufficient for drug candidates intended for long term use, provided the candidate is studied in rats or other appropriate species to evaluate the potential for tumor production. Wherever possible the proposed clinical formulation should be used in preclinical sub-chronic and chronic toxicology evaluation, because formulation excipients can be important in the extent and duration of delivery and in local tolerance. Dose Selection for subchronic and chronic toxicology studies should be based on the results from acute toxicity studies and pharmacokinetic evaluations. The three dose levels generally selected are 1. a no toxic effect level, which will be at least equivalent to or a multiple of the proposed human dose, 2. a dose level that produces a toxic effect in clinical observations, clinical pathology or histopathological changes and 3. a dose level between these two.

2. **Safety Pharmacology:** Safety pharmacological studies are those studies that investigate the potential undesirable pharmacodynamic effects of a

substance on physiological functions in relation to its exposure in therapeut range and above. These studies mainly involve assessing the effects of drug candidate for pharmacological activity on the functions of org systems other than the target system.

The Objectives of Safety Pharmacology Studies Includes:

1. Identifying undesirable pharmacodynamic properties of substance th may have relevance to its human safety.
2. Evaluating adverse pharmacodynamic and pathophysiological effec of the test compound that is observed in toxicological and/or clinic studies.
3. Investigating mechanism of the adverse pharmacodynamic effec observed and / or suspected.

The design of safety pharmacological study should include a size of grou that will allow meaningful scientific interpretation of the data generate The route of administration of the test compound should be the same expected clinical route of administration. Assessment of the effects b more than one route has to be performed if the test substance is intende for clinical use by more than one route of administration. eg. oral a parenteral route.

The organ systems that are essentially evaluated include central nervo system, cardiovascular system, and respiratory system. Additional org systems such as renal, autonomic nervous system and gastrointestin system can also be evaluated. Specific essential assessment criteria a supplemental studies that are required for these studies are tabulated belo

Safety pharmacology studies are not necessary for locally applied agen (Dermal or Ocular), if the pharmacology of the test substance is we characterized, and where systemic exposure or distribution to other orga or tissues is low. In case of cytotoxic agents, safety pharmacology studi prior to the first administration of human dose may not be require However, for cytotoxic agents with novel mechanism of action, conductir these studies may be useful. For biotechnology derived products th achieve highly specific receptor targeting, it is often sufficient to evalua safety pharmacology end points as a part of toxicology and/ pharmacodynamic studies. But for a product representing a nov therapeutic class and/or those products that do not achieve highly specif receptor targeting, a more extensive evaluation by safety pharmacolog studies are needed.

Safety pharmacology core test battery

System	Core Assessments	Follow up studies Assessments
'entral ervous ystem	Motor activity, behavioural changes, body temperature Ex. Modified Irwins test	Behavioural pharmacology, Learning and memory, ligand specific binding, Neurochemistry, visual, auditory and/or electrophisiology examinations.
'ardio-ascular ystem	Blood pressure, Heart rate, Electrocardiogram	Cardiac output, Ventricular contractility, vascular resistance
:espiratory ystem	Respiratory rate and other measures of respiratory functions eg: Tidal volume, oxygen saturation.	Airway resistance, pulmonary arterial pressure, blood gases, blood pH.

Supplemental Safety Pharmacology

Organ system	Assessments
Renal/ Urinary system	Renal Parameters Eg: Urinary volume, specific gravity, osmolality, pH, Fluid/Electrolyte balance, proteins, cytology and blood chemistry (Blood urea, Creatinine, plasma proteins)
Autonomic nervous system	Binding to receptors, functional responses to agonists and antagonists *in vitro* or *in vivo*.
Gastrointestinal system	Gastric secretion, gastrointestinal injury potential, bile secretion, transit time *in vivo*, ileal contraction, gastric pH measurement.

3. **Toxicokinetic Studies:** If the potential drug candidate is poorly absorbed, and has variable absorption or shows saturable absorption with increasing dose levels, administered dose has shown to be a non-uniform predictor of toxicity. It is important that the extent and duration of exposure is not changed after multiple dose administration. Thus, an ICH guide has been issued on the generation of toxicokinetic data to support the development of drug candidates.

   Objectives of Toxicokinetic study includes:

   The primary objective of toxicokinetic study is to describe the systemic exposure achieved in animals and its relationship to dose level and the time course of the study.

Secondary Objectives Include

1. Correlating the extent of exposure in toxicity studies to toxicologic findings and contributing to the assessment of the relevance of thes findings to clinical safety.
2. Supporting the choice of test species and treatment regimens that ar selected in non-clinical toxicity studies.
3. Providing information on the appropriate design for subsequent no clinical toxicity studies and human clinical studies, in conjunction wit toxicity results.

Toxicokinetic data, mainly focuses on the kinetics of a new therapeuti agent under the conditions of toxicity studies themselves, assisting in th interpretation of toxicity results, rather than characterizing the basi pharmacokinetic parameters of the drug candidate being studied. Thus toxicokinetics is an integral part of drug developmental studies and can b obtained from the test species in a toxicological study or in speciall designed supportive studies. These studies enhance the value o toxicological data generated, both in terms of understanding the results c toxicity tests and its comparison with clinical data as part of assessmer of risk and safety in humans. All toxicological studies need not include toxicokinetic component,. iIf the pharmacokinetic data for a dose leve range to be used in toxicity study has been generated then addition toxicokinetic evaluations are not generally necessary. Toxicology studies which are usefullyessentially supported by toxicokinetics, include singl dose studies, for which results from preliminary pharmacokinetics studie may be applicable, multiple dose studies in which toxicokinetic data ma predict whether multiple dose pharmacokinetics studies are necessary reproductive studies which may use different test species that have a altered absorption and disposition profile due to pregnancy, an carcinogenicity studies, in which the test species may be dosed differentl from other toxicology studies and changes in exposure can occur becaus of age.

4. Immunotoxicity Studies

Potential drug candidates have to be evaluated for possible immunotoxi effects. Toxicity to the immune system will be accompanied by variety o adverse events that mainly include the suppression or enhancement c immune response. Suppression of immune response leads to a decrease host defense mechanism making the individual prone to attack by infectiou agents, on the other hand, enhancement in immune response wil exaggerate autoimmune disease or mediates a hypersensitivity reaction

A standard toxicity study has to be conducted and specific signs for immunotoxicity have to be evaluated. Fig. 3.3 gives an overview of standard toxicity study and specific component that has to be looked into for signs of immunotoxicity.

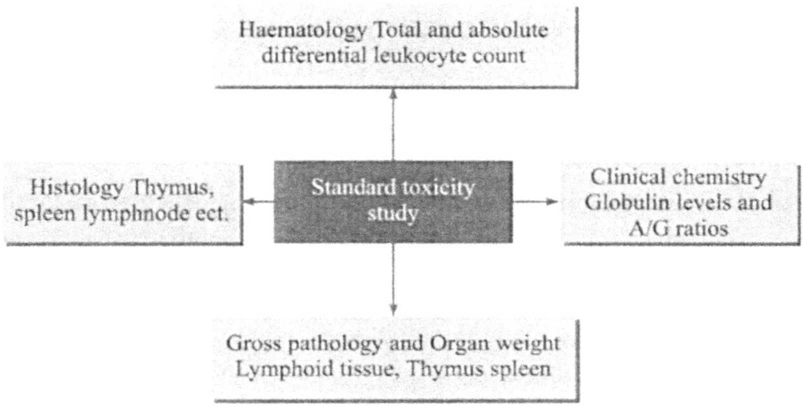

Fig. 3.3

1. **Haematology and Clinical Chemistry:** Total leukocyte counts and absolute differential leukocyte counts are generally performed to assess immunotoxicity. Changes in serum globulins gives an indication of a possible change in serum immunoglobulins. While evaluating globulin levels other factors like liver toxicity or nephrotoxicity hasve to be considered.

2. **Gross Pathology and Organ Weights:** All lymphoid tissue should be evaluated for changes at necroscopy. Organs like spleen and thymus must be weighed accurately.

3. **Histopathological Examination:** Histopathological changes in spleen and thymus must be evaluated. The changes may give an indication of systemic immunotoxicity. Lymphoid tissue that drains or contacts the site of drug administration should also be evaluated for changes.

Additional Immunotoxicity Studies

Following are the six factors that have to be considered and might prompt additional immunotoxicity studies.

(i) ***Findings from standard toxicity studies:*** Results from the standard toxicity studies should be thoroughly evaluated for any signs of immunotoxicity. These may include but not limited to Haematolgical changes (leukocytopenia, granulocytopenia, lymphopenia), alteration in immune system organ weights and histology (thymus, spleen, lymphnode), changes in serum globulins, increased incidence of infections or increased incidence of tumors.

(ii) ***Pharmacological Properties:*** If the pharmacological properties of the compound indicated that it has the potential to affect immune function additional immunotoxicity studies hasve to be carried out.

(iii) ***Intended patient population:*** Additional immunotoxicity studies for a compound must be performed if it is intended for use in immunocompromised patients by a disease state or due to concurrent therapy.

(iv) ***Structural similarity:*** Compounds that are structurally similar to compounds known to have immunosuppressive properties additional immunotoxicity studies have to be considered.

(v) ***Disposition of the drug:*** If the compound and or its metabolites are retained at high concentrations in cells of the immune system, additional immunotoxicity testing should be considered.

(vi) ***Signs observed in clinical trials or clinical use:*** If clinical findings are suggestive of immunotoxicity in patients exposed to the drug, additional non clinical immunotoxicity testing have to be carried out.

Additional immunotoxicity testing assays include, T-cell dependent Antibody response (TDAR), Immunophenotyping, Natural Killer Cell activity assays, Host resistance studies, Macrophage/ Neutrophil function assay and Aassays to measure cell-mediated immunity.

5. **Genotoxicity Studies:** An important aspect of preclinical safety evaluation is to screen potential drug candidates for their possible genotoxic effects. Standard genotoxic tests are *in vitro* and *in vivo* tests designed to detect compounds that induce genetic damage directly or indirectly by various mechanisms. Compounds that show positive results in these genotoxic tests have the potential to be human carcinogens/mutagens that is, they may induce cancer and or heritable changes.

It has been seen that no single tests is capable of detecting all relevant genotoxic agents, therefore a standard battery of *in vitro* and *in vivo* tests are carried out to evaluate genotoxicity. These standard tests are complimentary to each other rather than representing different levels of hierarchy.

Three tests are generally performed to screen candidate drugs for genotoxic potential.

1. A test for gene mutation in bacteria
2. An *in vitro* tests which involves cytogenetic evaluation of chromosomal damage by using mammalian cells such as human lymphoblastoid TK6, V79 and AS52 cells or an *in vitro* mouse lymphoma tK assay, an example for this is performing *in vitro* tests using L 5178Y cell lines.
3. An *in vivo* tests for chromosomal damage using haemopoietic cells.

Other tests like measurement of DNA adducts, DNA strand breaks, DNA repair or recombination, in addition to the above three standard tests can also be performed.

Generally drug candidates that give negative results in three standard genotoxic tests mentioned above are considered to have demonstrated an absence of genotoxic potential. For compounds giving positive results, based on their therapeutic use, may need to be tested more extensively. In some cases additional genotoxicity studies hasve to be carried out in appropriate models if the potential drug candidate has shown some effects in carcinogenicity bioassay with no clear evidence for a non-genotoxic mechanism, inspite, of the three standard tests results being negative.

6. **Carcinogenicity Study:** Carcinogenicity studies are performed to identify a tumorigenic potential in animals and to assess the relevant risk in humans. These studies usually involve most of the life span of the test species. Carcinogenicity studies hasve to be performed for any pharmaceutical compound which is expected to be used clinically for a continuous period of six months. The studies are generally conducted along with phase III human trials, but should be initiated much earlier in the drug development process if,

    1. The potential drug candidate's structure activity relationship suggests a carcinogenic risk.
    2. The drug candidate is a member of therapeutic class, which has shown a positive carcinogenic response.
    3. Repeated dose toxicity studies have shown evidence of preneoplastic lesions.
    4. Accumulation of drug or its metabolite in organ systems.
    5. Mutagenicity tests suggest that it may be a potential carcinogen.

    Mice and rats having a life span of approximately 18 and 24 months respectively are generally used in carcinogenic studies mainly because of their susceptibility to tumour induction and availability of large database on their physiology and pathology. ICH guidelines suggest that one rodent (rat) carcinogenicity study combined with another study usually a chronic or subchronic *in vivo* rodent study is sufficient to ascertain the carcinogenicity of potential drug candidate.

    The route of drug administration in test species should be the same as the clinical route of administration, wherever possible. An alternate route may be used if it gives similar metabolism and systemic exposure particularly to relevant organs, example: lungs for inhalation agents. Dose selection for carcinogenicity studies is an important parameter.

As per ICH guide lines the selected dose should

1. Provide the test species an exposure of the potential drug candida that allows an adequate margin of safety over the human therapeu exposure.
2. Be tolerated without significant chronic physiological functi impairment and are compatible with good survival.
3. Be guided by comprehensive animal and human data that focus on t properties of the drug candidate and the suitability of the test speci
4. Permit dose interpretation in relation to proposed clinical use.

The approaches that may be employed for dose selection includes, toxici based endpoints (MTD), pharmacokinetics endpoint, saturation of absorptio Pharmacodynamics end points and maximum feasible dose.

3.7 Developmental and Reproductive Toxicology

Developmental and reproductive toxicology studies are conducted to evalua any potential harmful effects of a drug candidate or its metabolite on mammali reproduction and to ascertain the potential risk to humans. These studies evalua male and female fertility, embryo and foetus, parturition and the newborn, t lactation process, care of the young and the potential teratogenicity of the dr candidate.

As per ICH guideline the combination of studies that are selected shou allow exposure of mature adults and all stages of development. To allow detecti of immediate and latent effects of exposure, observations are to be continu through one complete life cycle i.e. from conception in one generation conception in the following generation. For convenience of testing this integrat sequence is subdivided into following sequences:

- Premating to conception (Adult male and female reproductive functio development and maturation of gametes, mating behaviour, fertilizatio
- Conception to implantation (adult female reproductive functio preimplantation development, implantation)
- Implantation to closure of the hard palate (adult female reproducti functions, embryonic development, major organ function).
- Closure of the hard palate to the end of pregnancy (adult fema reproductive functions, foetal development and growth, organ developme and growth)
- Birth to weaning (adult female reproductive functions, neonate adaptati to extrauterine life, preweaning development and growth)
- Weaning to sexual maturity (post weaning development and grow adaptation to independent life, attainment of full sexual functions)

The animals used in these studies must be well defined with respect to their ealth, fertility, fecundity, prevalence of abnormalities, embryo fetal deaths, and e consistency they display form study to study. Generally the animals that are sed are young mature adults at the time of mating with the females being rgin. These studies are carried out in mammals, and it is desirable to use the me species and strain as in other toxicological studies. As a common practice, udies involving sequences A, B, E and F are combined into a single study and e conducted in rats, where as studies of sequence C and D are conducted parately in rats and rabbits.

As with most toxicology studies, three dose levels are used in developmental nd reproductive toxicology studies. Using similar doses in these studies as in peated dose toxicity studies will allow interpretation of any potential effects n fertility in context with general systemic toxicity. Commonly a dose finding udy, which involves toxicokinetic evaluation, is also conducted in pregnant nimals.

Studies of sequences A and B are designed to evaluate fertility and general productive performance, uses sexually mature male and female rats. Male rtility is determined by premating dosing of atleast 4 weeks and with dosing ontinuing throughout the mating period. Histopathology of the testes and sperm nalysis is carried out to detect effects on spermatogenesis. Female fertility is etermined by premating dosing of atleast 14 days with dosing continuing during e mating period. Copulation is evaluated daily by vaginal smears or by bservation of the copulatory plug. Day 0 of gestation is when proof of copulation discovered. Half of the females are sacrificed at a point after midpregnancy, sually day 13 of gestation and are examined for the number and distribution of mbryos in each uterine horn, embryos undergoing resorption, and the presence f empty implantation sites. The other half of the females are allowed to deliver ormally, and the litter size, number per litter alive or dead, and any abnormal bservations during gross examination are noted.

The studies of sequence C and D are designed to ascertain the potential mbryotoxic and teratogenic effects of the drug candidate. These studies are enerally conducted in a rodent (rat) and non rodent (Rabbit). The drug candidate administered during the period of organogenesis, which is usually considered estation day 6 to 15 for mice and rats and gestation day 6 to 18 for rabbits. etuses are delivered by cesarean section a day or two before anticipated arturition. Soft tissue alterations and skeletal abnormalities are evaluated in the tuses. In case of rats half of the fetuses are examined for visceral alterations nd the other half or skeletal abnormalities, whereas for rabbits micro dissection chniques for soft tissue alterations allow all the fetuses to be examined for oth soft tissue and skeletal abnormalities.

Studies in sequences E and F are designed to evaluate effects on perinat and post natal development of pups and on maternal functions. The drug candida is administered to the rats from implantation to the end of lactation. At the tin of weaning, normally one male and one female offspring per litter are selecte for rearing to adulthood and mating to assess reproductive competence. Thes offspring are also evaluated by use of behavioral and other functional tests fo the study of physical development, sensory functions and reflexes and behaviou

Summary of the Principles of Toxicity Studies (The FOUR Rs)

It has been thus suggested that the modern approach to toxicological testin should be based on the principle of four "R"s - 1. Reduction, 2. Refineme 3. Replacement and 4. Responsibility.

1. **Reduction:** The number of animals used should be reduced to the absolu minimum, which will be necessary to give a acceptable result. This ca be achieved by placing greater focus on the objectives of the study, bette experimental designs and minimizing the need for repeat studie Reassessing lethal tests (LD_{50}), using prescreening, using *in vitro* tes where possible and promoting dissemination and greater sharing of te data worldwide, will contribute to a large extent in reduction of anim testing. Wherever possible, reusing animals or using them for multip tests (for example, ocular and dermal) can also be considered.
2. **Refinement:** Adjustments to study designs and techniques have led to reduction in the number of animals used and in many cases testing hav become more humane, without compromising the scientific validity and i some cases improving it. Examples of such refinements include, fixe dose scheme adopted by the British toxicological society, limit tests, an modification to eye irritation tests.
3. **Replacement:** A number of alternate tests have been studied, an significant progress has been made in developing *in vitro* replaceme models. These currently include eye, muscle, and skin irritatio phototoxicity, photosensitization and target organ toxicity testing.
4. **Responsibility:** It is the responsibility of the toxicologists and th researchers to reduce animal suffering and to treat the animals in the be possible way.

3.8 Conclusion

The process of preclinical drug discovery normally proceeds linearly from, th drug discovery research and the pharmacology in which the lead candidate i evaluated in animal models to show that the compound has the potential c mediating a human disease, to the next step involving develop ability assessmen which provide preliminary data on safety, pharmacokinetics, drug delivery an xicity of the lead candidate and finally to the preclinical evaluation that mainly cuses on the safety aspects of the potential drug candidate. Careful design, nduct and interpretation of the results from these non-clinical research periments normally determine, which discovery lead haves the necessary tributes to become human therapeutic products for humans. These studies can used to "weed out" those candidates that have unacceptable pharmacology toxicology profiles before or shortly after the initiation of human clinical trials, ere by saving both time and money.

uggested Reading

1. Guo Y, Shafer S, Weller P, Usuka J, Petiz G. Pharmacogenomics and drug development. Pharmacogenomics 2005 (8): 857-64.
2. Drug Discovery and Evaluation – Pharmacological assays. H.Gerhard Vogel (ed). Springer publication
3. New Drug Approval Process. Ed. Richard A Guarino. Marcel Dekker Publication.

CHAPTER 4

Phases of Clinical Trials

A.A. Mateen, Y. Sridhar, M.U.R Naidu and P. Usha Rani

4.1 Introduction

Development and testing of drugs demands high standard of scientific rigor and safety monitoring because of the unique ethical issues of using humans as research subjects and patients who will eventually receive the treatment.

The primary goal of a clinical trial is to establish cause and effect. In order to reach this goal, the study must isolate the effect of a treatment and rule out factors that could lead to misleading findings. This may include the elimination or at least minimization of *bias*, *confounding* (where the estimate of the treatment effect is distorted due to the impact of another influential variable), and uncertainity. A clinical trial is a *prospective study* (a study that is planned in advance in which data is collected, as opposed to using data that has already been collected in a retrospective study), which permits manipulation and measurement of a treatment in a controlled setting and thus provides the optimal strategy to avoid bias and confounding, minimize randomness, and quantify the uncertainty in the study outcome. If done properly, it is highly likely that the results will be considered valid.

After an experimental investigational product has been shown to be the potential candidate for its therapeutic usage based on laboratory and animal studies, an *Investigational New Drug* (IND) application is filed with the

;gulatory authorities (For example, DCGI in India or FDA in US), requesting approval to begin experimentation in human subjects. If the IND application ıpproved, then a series of phases of clinical trials are planned, that progressively amine the optimal dose, safety, short-term and long-term effects on the body, d ultimately the effectiveness of an experimental therapy in humans. Usually : drug is studied in four different phases. Phase I to Phase III trials are nducted before the formal approval from the regulatory authorities to market : drug is obtained. Based on the results of Phase I, II and III, an application r "New Drug" is filed with the regulatory authorities by the sponsor, and once proved the drug is marketed. Following the marketing of the drug, Phase IV ıdies are conducted. Fig. 4.1 summarizes the various phases of clinical trials d the time required for each phase. In this chapter we will discuss the various ıases of clinical trials and their importance in drug development.

	Discovery/ Preclinical Testing		Phase I (Clinical trials)	Phase II (Clinical trials)	Phase III (Clinical trials)		Phase IV	
Year	6.5	File IND at FDA	1.5	2	35	File NDA/BLA at FDA	1.5	Additional post marketing testing required by FDA
Test Population	Laboratory and animal studies		20 to 100 healthy volunteers	100 to 500 patient volunteers	1,000 to 5,000 patient volunteers			
Purpose	Assess safety, biological activity and formulations		Determine safety and dosage	Evaluate effectiveness, look for side effects	Confirm effectiveness, monitor adverse reactions from long-term use		Review process/ approval	
Success rate	5,000 compound evaluated		5 enter trials				1 Approved	

Fig. 4.1

2 Phase I

ıase I trials are the first-stage of testing in human subjects. Normally these als are done on small number of healthy subjects, usually 20-80, divided in ıall groups known as "Cohorts". This phase includes trials designed to assess e safety, tolerability, pharmacokinetics and pharmacodynamics of an perimental product. These trials are almost always conducted in specialized inical pharmacology units, where the subject can be observed by full-time edical staff. The subject is usually observed till several half-lives (7-8) of the ug have passed. Phase I trials also normally include dose-ranging studies so at, doses for clinical use can be refined. The tested range of doses will usually : a small fraction of the dose that causes harm in animal testing. Although ıase I trials most often include healthy volunteers, there are some circumstances hen patients are included, such as in oncology (cancer) and HIV drug trials. In ıase I trials of new anticancer drugs, for example, patients with advanced (metastatic) cancer and those who are refractory to standard therapy are enrolle These trials are usually offered to patients who have had other types of thera and who have few, if any, other treatment choices. The objectives of Phase trials include, but are not limited to assessing the safety and tolerability of t new experimental product.

(a) Estimate the maximum tolerated dose (MTD).

(b) Identifying the Pharmacokinetic profile of the new product.

(c) Standardizing the dose through dose escalation studies a recommendation of a dose for further studies.

(d) Helping in proper designing of Phase II and Phase III trials.

Dose Calculation and Maximum Recommended Starting Dose

In 16th century, Paracelsus was credited with the observation that "All substanc are poisons. There is none that is not a poison. Only the dose distinguishes poison from a remedy". Even the most innocuous substance can be harmful an improper dose or via wrong route of administration. Therefore, with a New Chemical Entity (NCE) that is administered to human subjects for the fi time, it is necessary to define a dose that poses as low a risk as possible. Do justification is the cornerstone of a first in human trial of a NCE and a considerab amount of work goes into defining a safe starting dose for human volunteers.

Different approaches have been used to arrive at the safe starting dos Guidance by the center for drug evaluation and review (CDER) provides lucid, well articulated algorithm for defining the safe starting dose for first human studies. Here we briefly summarize the guidance that will help in arrivi at the starting dose for first in human studies.

From the preclinical toxicity studies two key parameters are identified: (the no observed adverse event level (NOAEL) and (b) the maximum tolerate dose (MTD). The maximum tolerated dose is the highest dose that can be give without killing and the dose above which substantial toxicity is observed. MT helps in identifying the target tissues and other toxic results that may be expecte with the compound. The NOAEL is the level at which no overt toxicity is observe either clinically or histopathologically and is the most important criteria establishing the safe starting dose from preclinical studies.

Calculating the Human Equivalent Dose (HED) and MSRD

Once the NOAEL is determined in the most metabolically appropriate a sensitive species, it should be converted to a dose that is scaled to body surfa area. This scaled dose is referred to as human equivalent dose (HED). Usual most toxicology studies define doses in terms of mg/kg of body weigh extrapolation to humans is given in mg/m^2. It has been seen that, especially f certain anticancer drugs, the MTD in rodents and humans correlated best wi each other when doses were expressed in terms of body surface area. Conversic

' the animal dose from mg/kg to a dose in mg/m^2 can be performed by using andard conversion factor (Km). Table 4.1 gives the Km value for different ecies.

Table 4.1 Conversion of Animal doses to Human Equivalent Doses (HED) based on body surface area.

Species	To convert animal dose in mg/kg to dose in mg/m^2 multiply by Km below
Human	37
Mouse	3
Hamster	5
Rat	6
Guinea pig	8
Rabbit	12
Dog	20
Monkey	12

For example to convert 20mg/kg NOAEL dose in dog to a dose in mg/m^2, it as to be multiplied by 20 (Km factor of dog), multiplying 20 by 20 the result ould be 400mg/m^2. To determine the equivalent human dose in mg/kg, the sult has to be divided by human Km factor which is 37, this results in HED of 0.8 mg/kg. It can be seen that when expressed in terms of body surface area, e NOAEL dose level in dog of 20mg/kg yields a HED of 10.8 mg/kg. This pproach provides a more conservative estimate than the one obtained by simply pplying the animal-derived NOAEL dose on a mg/kg basis to human (In this xample, the dose is approximately reduced by half). The final step in the dose etting process is to adjust the HED downward by some multiplicative factors, ased upon likelihood that toxicology testing will fail to completely capture the pectrum and amplitude of human toxicity. In the absence of data that would ise concerns about potential effects on humans, the recommended safety factor . 10.0. Thus, if the HED of the most sensitive species tested is 10.8 mg/kg, the commended starting dose would be 1.08mg/kg (or about 70.2mg for a 65kg erson). The safety factor may be decreased if the compound is a member of a lass for which the toxicity in humans is known and in which the dose, route, hedule and duration are the same. However, for most NCEs with novel modes f action, this will not be the case. The safety factor can be increased in the vent where the other data are particularly worrisome, such as when the ioavailability is highly variable within and between species, making predictions bout human pharmacokinetics difficult. Also when animal data suggest that xicities are severe and cannot be monitored easily (Silent histopathological lesions), when effects are irreversible, and there are unexplained deaths in toxici study. Compounds that have a very steep dose-response curve for a given tox endpoint may also need additional safety margins to ensure adequate protectic of human subjects.

Single Dose Phase I Study

In single ascending dose phase I trials, subjects are grouped in small groups (three to six and are given a small dose of the investigational product and observe for a specific period of time. If they do not exhibit any adverse side effects, new group of subjects is then given a next higher dose. This is continued un intolerable side effects start showing up. At this dose level, the drug is said have reached the Maximum Tolerated Dose (MTD).

Multiple Dose Phase I Study

Multiple Ascending Dose studies are conducted to further understand th pharmacokinetics/pharmacodynamics of the drug. In these studies, a group (subjects are administered a low dose of the drug and the dose is subsequentl escalated up to a predetermined level. Different approaches will be employe for escalating the dose. Samples (blood, and other body fluids) are collected various time points and analyzed to understand how the drug is absorbe distributed, metabolized and excreted within the body.

Safety and Tolerability

The primary objective of a phase I trial is to provide first information on th safety and tolerability of a NCE in man and to evaluate the potential risk of th compound before giving to the target population. Safety and tolerability paramete should include subjective tolerability as well as parameters of objective tolerabilit Table 4.2 lists the basic criteria for the assessment of safety and tolerability i phase I studies.

Table 4.2

Criteria for assessment of safety and tolerability
Subjective tolerability
Clinical examination
Vital Signs
ECG parameters
Laboratory Parameters
• Complete haematology including Coagulation.
• Liver, Kidney, Pancreas
• Muscle enzyme
Specific organ safety

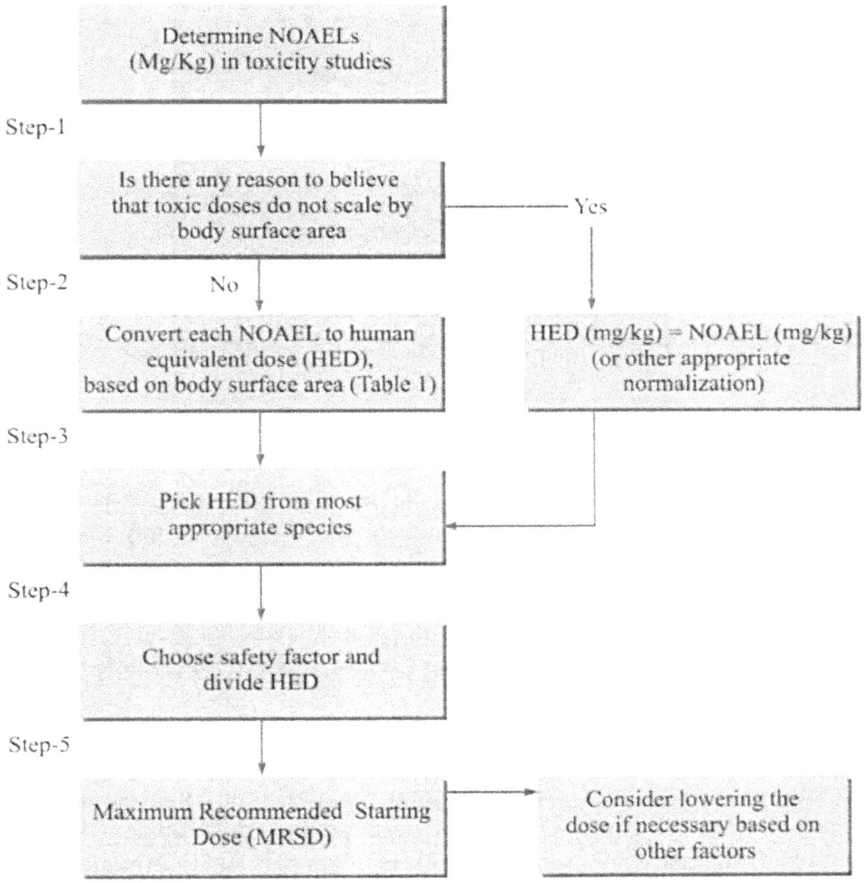

One example of a safety parameter, which is routinely measured in every hase I study, is the ECG parameter of the QT interval. The QT interval is a neasure of ventricular action potential duration and represents the time from he first ventricular activation to the latest ventricular repolarization. Drugs causing QT prolongation have been suggested to exert a direct or indirect membrane ffect and it has been linked to arrhythmias. Another example is increase in aboratory parameters of renal and hepatic function (blood urea, serum creatinine nd SGOT, SGPT, GGT and alkaline phosphatase).

harmacokinetics

nformation on basic pharmacokinetic variables can be obtained during phase I tudy. Table 4.3 enlists the pharmacokinetic parameters that are generally studied n phase I trials.

Table 4.3

Criteria for assessment of Pharmacokinetics in Phase I study
C_{max}, C_{min}, C_{av}, C_{ss} t_{max}
Volume at steady state (Vss), Vc
$t_{1/2}$
Ke
Clearance Cl (total, renal, extra-renal)
Accumulation, dose proportionality
Protein binding
Mean Retention Time (MRT)
Metabolism

The maximum plasma concentration and the inter-individual variability wit respect to the maximum individual exposure are of immense importance. Th safety assessment must be based on the comparison of safety results an exposure. Unwanted drug exposure in late clinical trials can be prevented b studying different routes and rates of administration, pharmacokinetic modellin screening for genetic polymorphism and specific risk groups (elderly).

IND Application for Phase I Trials

In many ways, the investigational new drug (IND) application is the result of successful preclinical development program. The IND is also the vehicle throug which a sponsor advances to the next stage of drug development known a clinical trials (human trials). Generally, this includes data and information in thre broad areas:

1. ***Animal Pharmacology and Toxicology Studies:*** Preclinical data to perm an assessment of whether the product is reasonably safe for initial testin in humans.

2. ***Manufacturing Information:*** Information pertaining to the compositio manufacture, stability, and controls used for manufacturing the dru substance and the drug product. This information is assessed to ensur that the company can adequately produce and supply consistent batche of the drug.

3. ***Clinical Protocols and Investigator Information:*** Detailed protoco for proposed clinical studies to assess whether the initial-phase trials wi expose subjects to unnecessary risks.

Also, information on the clinical investigators and the sites of the trial has t be provided to the regulatory authorities.

3 Phase II Studies

is phase of studies represents the first opportunity to evaluate the efficacy of drug on target patient population and is also an extension of safety evaluation ne previously on healthy volunteer population in phase I. Phase IIa is generally ilot trial involving a small number of patients (20-40), who are stable without mplications, and aims at establishing the proof of concept (POC) and gathering icacy data. In phase IIb, larger number of patients (150-500) are included metimes using multiple sites) and treated with the study drug in order to get re detailed information about efficacy, dose response and safety. This step ıst provide enough result to determine the design, dosage, objectives and thods to be used in phase III. Some study designs are now mixing phase I and by looking to validat pharmacodynamic markers in healthy subjects or by tegrating a small group of patients with mild disease (phase IIa) as an ancillary oup in phase I study. Biomarkers can provide great predictive value in early ug development if they reflect the mechanism of action for the intervention en if they do not become surrogate endpoints.

oof of Concept Studies

oof of concept trials are initial studies performed to find out whether the new ug developed is effective in its intended purpose. Proof of concept trials are ten conducted before formal dose ranging study are started and usually present the first time a drug is tested in intended patient population. The results these trials are crucial for sponsors to decide on the future prospects of the ug development plans. During drug development process it is necessary to d out failures quickly so that expensive phase Iib and III studies are not nducted on a drug candidate that is not safe and / or ineffective. Proof of ncept trials are crucial in exposing such failures. If the result from a proof of ncept study is positive, it can be used in designing subsequent studies and fining the remaining phase Iib and phase III trials. Proof of concept trials can performed as soon as the data and results from the appropriate toxicology ıdies and phase I trials are available and demonstrate that the drug is well lerated at the doses tested.

udy Design for POC (Proof of Concept) Studies

e objectives of proof of concept studies can be achieved by using different ıdy designs.

rallel group design: Two groups are matched for confounding variables and en designated a treatment, either a test or control, on which they remain for e duration of the study. This allows comparison of the experimental treatment oup versus a control group at multiple points and also helps in defining the armacokinetic parameters by drawing the area under the curve.

Cross-over designs: these are useful in minimizing variability in responses sin each patient serves as his/her own control by receiving each treatment in sequential fashion.

Uncontrolled study designs: These designs are appropriate in cases where t key objective is improvement in a condition or disease state over time, as comparison of changes to baseline can be analyzed.

Historical controls: if historical data on appropriate control groups are availab it is possible to compare responses in the Proof of concept trial to observatio from previous studies, either with comparative products or placebo. Howev these designs are least reliable as many factors can vary between the historic control design and the proof of concept study design. An example is the differen in the patient population.

Choice of Doses and End Points

Proof of concept studies are usually conducted using only one or two dose leve usually the dose that is found to be safe in phase I trials. Surrogate end poir are generally employed in proof of concept trials. For example, in an HIV tri decrease in viral load and an increase in CD4 and CD8 counts can be employ as surrogate markers. However, care should be taken to use validated endpoin to strengthen the application of the findings of Proof of concept trials for furth studies.

Incorporating the provision for measuring pharmacokinetic parameters in pro of concept study designs is a quick and efficient way to determine th bioavailability of the drug in patient population. Pharmacokinetics Pharmacodynamic (PK/PD) analysis is also very useful in assessing th relationship between effect and concentration to help in guiding further clinic development.

Positive results from a Proof of Concept trial will usually result in progressio of the drug to further development. The next step is to conduct additional effica studies in the patient population for which the drug is intended, in order to defi the specific dose that produces the best risk–benefit ratio. It is important th the correct dose–response studies are conducted during Phase Iib to establis an appropriate dose for Phase III studies.

Dose Response Study

Knowledge of the relationship among dose, drug concentration, drug concentratic in blood and clinical response (efficacy and undesirable effects) is important f the safe and effective use of drugs in individual patients. This can help in identifyir an appropriate starting dose, the best way to adjust dosage to the needs of particular patient and a dose beyond which, if increased would be unlikely provide added benefit or would produce unacceptable side effects. Assessme of dose response is an integral component of drug development process wi udies designed to assess dose response, an inherent part of establishing the fety and effectiveness of the drug. Conducting dose response studies at an early age of clinical development may reduce the number of failed Phase III trials, eeding the drug development process and conserving development resources.

udy Designs

ne choice of study design and patient population in dose response trials will ainly depend on the therapeutic indication under investigation and the severity the disease, in the patient population of interest. For example, the lack of propriate salvage therapy for life threatening or serious conditions with eversible results may ethically preclude conduct of studies at doses below the aximum tolerated dose. A homogenous patient population will generally allow hievement of study objectives with small number of subjects given each eatment. On the other hand, larger and more diverse population allows detection potentially important covariate effects. Dose response information is best otained from trials specifically designed to compare several doses. If the control oups are similar, a comparison of results from two or more controlled trials ith a single fixed dose might be useful. Some of the study designs that can be nployed are:

(a) ***Parallel Dose Response:*** Randomization to several fixed dose groups (the randomized parallel dose response study) is a simple design to evaluate dose response. The fixed dose is the final or maintenance dose, patients may be placed immediately on that dose or titrated gradually to it. In either case the final dose is maintained for a time adequate to allow the dose response comparison.

(b) ***Cross over Dose-Response:*** A randomized multiple cross over study of different doses can be successful if drug effect develops rapidly and efficacy variables return to baseline quickly after cessation of therapy. The advantages of this trial is that each individual receives several different doses so that the distribution of individual dose response curves may be estimated and also the number of participants needed will be less.

(c) ***Forced Titration:*** A forced titration study is one in which all participants move through a series of rising doses. It is similar in concept and limitations to a randomized cross over dose response study, although the assignment to dose levels is graded and not random.

(d) ***Optional Titration (Placebo Controlled Titration to End Point):*** In this design, patients are titrated until they reach a well characterized favorable or unfavorable response, defined by dosing rules expressed in the protocol. This approach is most applicable to conditions where the response is reasonably prompt and is not an irreversible event such as stroke or death.

Dose response data are desirable for almost all new chemical entities. The data should be derived from studies that are sound and scientifically based. variety of different designs, some of which discussed above, can give val information.

4.4 Phase III Trials

Phase III studies are generally double blind randomized controlled trials on lar patient groups (1000-3000 or more) and are aimed at being the definiti assessment of the efficacy of the new therapy. The goal of phase III trials is take a new treatment that has shown promising results when used to treat small number of patients with a particular disease and compare it with the curre standard of care for that specific disease. In this phase, data is gathered fro large numbers of patients to determine whether the new treatment is mo effective and possibly less toxic than the current standard treatment. Althou Phase III trials focus on patients with a specific disease, they typically inclu patients of various ages, multiple ethnicities, and both genders so that the resul once obtained, may be applicable to a large number of patient population. Pha III trials are the most expensive, time-consuming and difficult trials to desi and run, especially in therapies for chronic conditions. These trials take ma years to complete. Phase III trials usually consist of both short term and lo term safety and efficacy studies.

Short-Term Safety and Efficacy Trial

These trials are primarily aimed at evaluating the safety and efficacy of a ne therapy with a short duration of treatment and comparing its efficacy with t standard therapy. An example of this kind of trial would be a 24-week doub blind, placebo controlled randomized trial.

Long-Term Safety and Efficacy Trial

Regulatory authorities require that the results of long term safety and effica of a new treatment be submitted before the approval to market the drug given. These studies are aimed at evaluating the long-term safety and effica of a new treatment. The studies have to be conducted for a minimum period 2 years at multiple centers.

4.5 Phase IV Trials

Phase IV trials are conducted after the drug has been approved and market to provide additional details about the medicines efficacy or safety profil Different formulations, dosages, duration of treatment, medicine interaction a other drug comparisons may be evaluated. Different age groups, races and oth types of patients can also be studied. Detection and defining the previous unknown or inadequately quantified adverse reactions and related risk facto e an important aspect of Phase IV studies. These trials may observe the fects of the drug in the settings where it is used in practice. These trials may ct as surveillance of safety, and/or may be used as a marketing tool to expand e use of the drug to new populations. Phase IV trial is conducted to find out:

- More about safety and side effects of the drug.
- Long-term risks and benefits.
- Efficacy of the drug when it is used more widely than in clinical trials.

Usually Phase IV studies consist of Comparative studies, studies involving vailable medication tested for new indications, or trials of new dosage and ormulation of approved medication. The term post marketing surveillance is sed to describe those clinical in Phase IV (i.e. the period following marketing) at are primarily observational or non experimental in nature, to distinguish them om well controlled Phase IV clinical trials or marketing studies.

uggested Reading

1. Spilker B. Guide to clinical trials. New York 1991. Raven Press
2. Spriet A, Dupin Spriet T, Simon P. Methodology of clinical trials. 2nd ed. Basel 1994; Karger.
3. Storer BE. Design and analysis of Phase I clinical trials. Biometrics 1989; 45: pg925-37.
4. John O Grady. Hand book of Phase I/II clinical drug trials. CRC Publication

CHAPTER 5

Phase 0 - Micro-dosing Studies

P. Usha Rani and M.U.R. Naidu

5.1 Introduction

Human drug development is a dynamic process that keeps pace with the recent advances within the pharmaceutical and analytical labs, combined with a precise understanding of the integrated pharmacokinetic-pharmacodynamic-pharmacogenomic drug profile. Sub-optimal pharmacokinetics is stated to be a reason for failures during drug development. Too low concentrations of drug at the target organ for shorter time can lead to efficacy failures while wrong concentrations, reaching wrong targets for longer time may lead to toxicity. Thus a new experimental approach has been developed known as ***Phase 0 or micro-dosing*** studies to address issues pertaining to drug metabolism and pharacokinetics. Micro-dosing methodology is referred to as a new viable "tool" in the drug development "toolbox". In Micro-dosing, extremely low, nonpharmacologically active doses of a drug are used to define the agent's pharmacokinetic profile in humans. Thus, by definition, micro-dosing means use of " less than $1/100^{th}$ of the dose calculated to yield a pharmacological effect of the test substance to a maximum dose of <100 microgrms (European Medicines Agency). However, in addition to this, US FDA suggests a maximum micro-dose of <30 nanomoles for protein products.

2 Micro-dosing and Exploratory-IND Studies

ploratory IND studies are intended to provide clinical information for a new ıg candidate at an earlier phase of drug development and help to identify itable candidates for continued development. These studies also provide portant information on pharmacokinetics and bioavailability of a candidate ıg. Thus micro-dosing and exploratory IND services help to

1. Obtain early human pharmacokinetics;
2. Select the best candidate from multiple preclinical candidates
3. Add value to drug pipeline with early clinical data.

From preliminary toxicology data of animals, pharmacokinetic data for initial se selection can be obtained in a short time of 4-6 months whereas in Phase I ıdies it takes 12-18 months. During drug development, when a large number me-too compounds are screened and found to have similar or differing animal armacokinetics, comparative human micro-dose studies can be done to establish armacokinetics. This PK data can further be used to

1. Help in selection of the ideal candidate drug,
2. Obtain the first in man dose of that selected drug
3. Establish the tentative pharmacological dose and
4. Calculate the probable cost of the deliverable.

A further question arises which states whether it is ethical to expose a subject a pharmacological dose unnecessarily when later, due to poor pharmacokinetics e development of the drug will be halted.

3 Characteristics of an Ideal Candidate Drug for Phase 0-Micro-dosing studies

ı ideal candidate drug for Phase 0 studies is one for which all the following aracteristics are present

(a) Primarily pharmacodynamic endpoints important for further successful clinical development
(b) Credible biomarkers or targets present
(c) Drug is expected to have a wide therapeutic window
(d) Target or biomarker modulation is anticipated at nontoxic doses and over short durations of exposure (eg $\leq$7 days)
(e) Smaller samples size of about 10-15 subjects is required to determine target or biomarker modulation

5.4 Goals of Microdosing Studies

The main goal of phase 0 trials is to obtain information from relatively sma group of subjects receiving nontoxic doses of drug, which would help in desig and potential success of subsequent larger phase I and II trials. Phase 0 tri involves a rational transition from preclinical to clinical development and hen requires close collaboration between laboratory, drug development and clinic scientists.

The goals of phase 0 include the following:

(a) Assess whether the mechanism of action defined in preclinical studies achieved in humans
(b) Define specific biomarkers or targets in human studies
(c) Determine special probes and methods to assess the pharmacokineti of the drug
(d) Develop novel models to evaluate the pharmacodynamics
(e) Define a human PK-PD relationship data for an agent prior to phase testing

The goals and design of phase 0 differs from traditional phase I trials as see from the Table below

Features	Phase 0 studies	Phase I studies
Basis for starting dose	Results from limited preclinical toxicity studies to support exploratory IND	Results from full standard IND-directed preclinical studies
Primary end point	Establish a dose range that modulated targets, which can be used in further drug development studies	Establish dose limiting toxicities and maximum tolerated dose
Amount of dug required and duration of dosing	Limited dosing usually single and less than 100 micrograms required	Single and multiple dosing an about 100 grams of drug required
Pharmacokinetic analysis	Real time. Special requirements like C14 labeled compounds if using AMS	Normally there are no specia requirements Samples are batched usually and analyzed at a later point.
Regulatory requirements	Firmly established	Limited and not clear

.5 Technical Advances

has been hypothesized that micro-dosing will help to reduce or replace the ctensive animal testing of compounds for kinetics which may later be rejected human studies. Thus micro-dose studies use minute quantities of drug and are ot intended to produce any pharmacologic effect when administered to humans nd therefore may not cause any adverse events also, but may produce useful harmacokinetic information and help in further development of the compound. owever, ultrasensitive and specific analytical methods capable of measuring ug and metabolite concentrations in the low picogram to fentogram range are quired. Most common of these are liquid chromatography coupled with tandem ass spectrometry (LC-MS-MS) and Accelerator Mass Spectrometry (AMS). he latter method is extremely sensitive compared to LC-MS-MS, however, ompounds must be isotopically labeled mostly with C^{14}. This may be a isadvantage to the use of AMS but, dose of radio labeled compound use for uman studies is as low as 100 nanocuries.

.6 Regulatory Issues

ccurate characterization of kinetics of a drug over time after administration is n important regulatory requirement. This can be achieved by administration of adio labeled drug to the subject and following its fate in plasma and excreta. ince AMS studies require very minute quantities of radio labeled compounds it ay not be a significant source of maternal risk by regulatory authorities. urthermore, recently, EMEA and FDA have published articles and supported ich evaluation.

.7 Patient's Perspective

rom a patient's perspective, the micro-dosing concept provides an opportunity help improve how some drugs can be evaluated. Their involvement can help searchers and future patients identify possible treatment paths in a fast pace as to either discard drugs that will not work or promote those that show tivity. The limited study drug exposure and risk of potential interventions are al challenges to participation. Patient risk and the risk-benefit ratio of phase 0 ial is a major consideration. Further, more there is no chance of direct medical enefit to subjects enrolled in phase 0 trials. To consider Micro-dosing studies hical, they must minimize risk to participants while offering important scientific alue to society.

In designing phase 0 trials, it is important to ensure that participation will not lversely affect a participant's eligibility to enroll in subsequent therapeutic trials adversely delay other therapies. Receiving a new drug as a part of micro-osing studies should not limit participants from enrolling in other studies with this agent or similar class of agents. Further, participants may not be required wait for longer washout period before staring other trials due to non-therapeut nature of phase 0 trial having minimum drug exposure. Thus a shorter washo period of 2 weeks or less may be sufficient.

5.8 Informed Consent Issues

Because of the risk to benefit ratio of micro-dosing studies, participan understanding is critical and their understanding becomes a priority. Hence, the is a clear and imperative need for investigators to carefully explain the differenc between experimental research and therapeutic care very objectively. Furth documenting that a participant fully understands the objectives, procedures, ris and benefits is important. Measurement of understanding is recommended assure an eligible participant has understood information given to him. It prudent to include documentation of understanding for early phase studies. Th process will be important as phase 0 trials are introduced and conducted study their unique informed consent process.

5.9 Single-dose Toxicity Study in Support of Micro-dosing Studies

1. Smaller toxicology package is required. As per the regulatory requireme animal studies at least in one species are required to establish micro-do in humans, but at a much reduced level. Further, if human screening compounds is done earlier in drug development process, fewer anim studies are required before Phase I clinical trials. Thus with compoun having unsuitable pharmacokinetic profiles, further animal studies can avoided.
2. Single dose exposure via intended clinical route
3. Establish dose inducing minimal toxic effect or a margin of safety of clinic dose (eg 100X).
4. Endpoints include - Body weight, clinical signs, chemistry, hematolo and histopathology.
5. No genetic toxicology required

5.10 Advantages of Micro-dosing

1. Micro-dosing requires minute quantities of the drug for safety testing. micro-dose is so small that when administered to human subjects, it is n intended to produce any pharmacologic action hence risk of adverse even is less.
2. *Save time:* Advance lead candidates to clinical development in mont not years.

3. *Save money:* Significantly reduced IND submission requirement and cost
4. Cost-effective approach to adding value to lead candidate or drug pipeline

.11 Limitations of Micro-dosing

'here are many questions which need to be answered regarding the predictive ccuracy of micro-dosing. We still do not have enough studies to clearly exemplify ,hether the body's reaction to a particular compound is similar when used as aicro-dose and in its pharmacological dose, otherwise, it could lead to false egatives (compound being rejected) or false positives (compound acceptable ased on micro-dose data but rejected subsequently when used in pharmacological oses).

Another question of concern is whether micro-dosing predicts pharmacokinetic arameters accurately for drugs showing nonlinear kinetics. This aspect has een addressed in CREAM study (Consortium for Resourcing and Evaluating MS Micro-dosing) where the disparity in warfarin disposition is described. 'hus caution should be exercised when micro-dosing is employed to drugs with omplex pharmacokinetics, especially during early drug development of new hemical entities. Thus a micro-dose may not be able to predict the behavior of clinical dose of the drug. However, though the CREAM study was not xhaustive, it demonstrated about 70% approximation between micro-dose and harmacological dose pharmacokinetics.

Another limitation of micro-dosing relates to metabolism and stability of certain ompounds. Some compounds dissolve readily at micro-dose yielding good bsorption characteristics, however, at therapeutic doses, they exhibit limited olubility and absorption becomes dependent on the rate and extent of dissolution vhich cannot be predicted at micro-dose levels.

Further, AMS (accelerator mass spectrometry) and PET (positron emission omography) are applied to analyse concentration of drugs in low picogram to emotgram range when micro-dose is used. These radiotracer assays have isadvantages of short tracer half-life and limited specificity (as assay may include netabolites also). Both for PET and AMS, drugs must be labeled at metabolically table sites.

.12 Conclusion

Aicro-dosing may later on become an accepted approach in drug development vhen first in man studies may begin with a Phase 0 study. However, true utility f Phase 0 micro-dosing studies lies with the ability to predict under what ircumstances this approach will provide data within a specified and acceptable ange as compared to therapeutic dose data. Capitalizing on the continuing rapid dvances in drug development technology, there is no question that decreasing the time of drug development, reduces the cost phenomenally. Until furthe information is available, we opine that micro-dosing strategy could complemer standard animal-to-human algometric scaling, redefining present phaseI stud designs. This strategy may help to reduce animal testing in identification of nove drug candidates.

Further, micro-dosing may help both patients and the pharma industry wit earlier availability of new test medication and reduced attrition of compounds a later stages of drug development.

Suggested Further Reading

1. Malcom Rowland. Micro-dosing and the 3Rs. NC3Rs #5 Feb 2006 pg 1-7
2. Colin Garner R and Graham Lappin. Commentary. The phase 0 micro dosing concept. BJCP 2006 ; 61 (4) : 367-370.
3. Joseph S.Bertino Jr, Howard E Greenberg and Michael D Reed. America College of Clinical Pharmacology Position Statement on the Use of Micro dosing in the Drug Development Process. J Clin Pharmacol 2007; 47 418-422.
4. Wood AJJ. A proposal for radical changes in the drug development process NEJM 2006; 355 : 618-623.
5. Dimasi JA. Risks in new drug development: approval success rates fo investigational drugs. CPT 2001; 69: 297-307.
6. Garner RC. Lesss is more: the human micro-dosing concept 2005; 10 (7) 449-451.
7. Martin Gutierrez and Deborah Collyar. Patient perspective on Phase clinical trials. Clin Cancer Rev 2008; 14 (12) pg 3989-91
8. Anthony J Murgo, Shivaani Kummar et al. Designing Phase 0 cance clinical trials. Clin Cancer Res 2008; 14 (12): 3675-3682.

CHAPTER 6

Clinical Research Planning

M.U.R. Naidu and P. Usha Rani

.1 Introduction to Clinical Trials

'he earliest evidence of clinical trial dates back to 20th May 1747, when in his lassical study James Lind evaluated six treatments in 12 patients of scurvy. ne of the two patients who was given lemons and oranges quickly recovered om signs and symptoms of scurvy and was fit for duty after 6 days. Earliest xperiments involved arbitrary non systematic schemes for patient assignment) various therapeutics, as described by Lind. Clinical trial has evolved into andard procedure in the evaluation of new drugs after World war II, which icluded use of control groups, random allocation of patients to control and xperimental groups and use of blind or masked assessment. It is only in the ast few decades, clinical trial has emerged as the preferred method in the valuation of medical intervention. Subsequently, several other comparative studies ave been conducted. Some of them are given in the following tables.

Clinical trial, which is properly planned and executed, is a very powerful xperimental technique for assessing the effectiveness of intervention.

Historical Events in the Development of Clinical Trials

Author	Period	Event
1747	James Lind	Experiment with untreated control group (Lind,1743)
1799	Haygarth	Use of sham procedure (Haygarth,1800)
1800	Waterhouse	U.S. Based small pox trial (Waterhouse,1800,1802)
1863	Gull	Use of Placebo treatment (Sutton,1965)
1923	Fisher	Application of randomization to experimentation (Fisher an Mackenzie year?)
1923	MRC Britain	Special committee on clinical trial created by the Medical Researc Council of Great Britain(Medical Research Council,1931)
1931	Amberson	Random allocation of treatment to groups of patients(Amberso et al,1931)
1937	NIH	Start of NIH grant support with creation of the National Canc Institute (National Institute of Health,1981b)
1944	PCTC	Publication of multicenter trial on treatment for common co (Patulin Clinical Trials committee,1944)
1962	Curran	Promulgation of Nuremburg Code for Human Experimentatio (Curran & Shapiro,1970)
1962	Hill	Publication of book on clinical trials (Hill,1962)
1962	Kefaiver & Harris	Amendments to the Food, Drug & Cosmetic Actof 1938 (Unite States Congress,1962)
1966	US PHSR	Publication of U.S. Public Health service regulation leading creation of Institutional Review Boards for research involvin humans (Levine,1981)
1967	Chalmers	Structure for separating the treatment monitoringand treatme administration process
1979	SCT	Establishment of Society for Clinical TrialsSociety for Clinic Trials.,Inc,1980
1980		First issue of Controlled Clinical Trials

Definition

A clinical trial is defined as a systematic and prospective study aimed to compa the effect and values of interventions against a control group in human subjec Clinical trial is a controlled experiment having a clinical event as an outco measure and done in a clinic or clinical setting and involving a person having specific disease or health condition (Meinert, 1996). An experiment is a seri of observations made under conditions controlled by the scientist, and a clinic trial is an experiment testing medical treatment on a human participant (Piantado 1997). A clinical trial must employ one or more intervention techniques li prophylactic, diagnostic, therapeutic devices, regimens and procedures, etc.

The design and conduct of any type of clinical trial requires three important considerations:

1. Study must examine valuable and important biomedical research questions.
2. It must be designed and based on a rigorous methodology that can answer a specific research question.
3. It must be based on a set of strong ethical principles to minimize the risk to trial participants.

The choice of an appropriate study design depends on the following factors and considerations:

1. Ability of the study design to answer the primary research question.
2. Whether the trial is studying a potential new treatment for a condition for which an established, effective treatment already exists.
3. Whether the disease for which new treatment is sought is severe or life threatening.
4. The probability and magnitude of risk to the participants.
5. The probability and magnitude of likely benefit to the participant.
6. The population to be studied – its size, availability and accessibility.

.2 Types of Clinical Trials

Clinical trials are usually classified into 4 Phases. (Detailed topic covered in chapter on – Phases of Clinical Trials).

Phase I trials are the earliest stage clinical trials used to study an experimental drug in human. They are typically small (less than 100 participants) and are often used to determine the toxicity and maximum safe dose of new drug. Phase trials provide an initial evaluation of drug safety and pharmacokinetics. Phase trials are commonly conducted with non-disease individuals (healthy volunteers). Some Phase I trials, that is, those studies for treatment of cancer and highly toxic drugs are performed in individual patients with advanced diseases who have failed on standard treatment.

Phase II trials are often aimed at gathering preliminary data on whether a drug has clinical efficacy and usually involve 100 to 300 participants. Frequently, Phase II trials are used to determine the efficacy and safety of an intervention participants with disease for which a new intervention is being developed.

Phase III trials are advanced stage clinical trials, designated to show conclusively, how well a new drug works. Phase III trials are usually larger, multi-institutional and typically involve from 100 – 1000 participants. They are comparative in nature with participants usually assigned by randomization to at least 2 arms, one of which serves as control, or a reference arm and one or more of which involve new intervention. Phase III trials generally measu whether a new intervention extends benefit or improves the health of t participants.

Phase IV trials are usually performed after a new drug or biological has be approved for use. These trials are post marketing surveillance studies aimed obtaining additional information about the risks, benefits and optimal use of intervention.

A much more inclusive and general approach that subsumes the fourth phas of clinical trial is suggested by Piantadosi (1979), who defines the fourth Phas as:

1. Early development study - testing the treatment mechanism (proof concept)
2. Middle development studies – treatment tolerability
3. Comparative – Pivotal confirmatory studies
4. Late development studies

6.3 The Planning Steps

Conducting successful clinical research projects requires careful planning befo the collection of data begins.

1. Develop good hypothesis for research.
2. Write the objectives and establish the priorities.
3. Define the variables needed.
4. Define the study population.
5. Refine the objectives into written, testable hypothesis.
6. Anticipate error and bias.
7. Develop the study design.
8. Estimate the sample size needed.
9. Write a research protocol.
10. Plan the data collection process.
11. Manage and Monitor data Collection
12. Manage the database
13. Analysis of observations and tabulation of results
14. Test the hypothesis with the correct statistical tests
15. Generate additional tables and graphs to explain the findings
16. Publication of results

tep 1: Develop Good Ideas for Research

The single most important step in successful clinical research is to come up with an important idea to study. Formulating important questions requires experience in the clinical field to be studied and interaction with other experienced investigators. Efforts that begin with methods or databases and proceed to questions rarely generate important conclusions. Instead, one should determine the important questions in the field and design the methods and databases to answer them. Once the idea has been thoroughly researched and discussed, it is often necessary to limit its scope before beginning the project. Often what seemed like a simple idea at first, mushrooms into a grandiose plan that is not feasible. The researcher must realistically assess the idea and narrow its scope to something that can be accomplished with the time and resources that might realistically be obtained.

tep 2: Write the Objectives and Establish Priorities

Early in the planning process, the researcher must make a list of all the objectives that are to be accomplished. When the objectives are not formally written, often there is an unrecognized confusion over how many objectives exist and what exactly are they. Once all of the objectives are identified, they must be rewritten in order of priority and relative importance. It is important to set the priorities among objectives before going further in the design.

tep 3: Define the Variables Needed

Once the objectives are formally written, the investigator develops a plan for accomplishing the objectives, that is, for answering the questions entailed by the objectives. Generally, the first decision in developing the plan is to enumerate the variables that will have to be measured to answer the questions. After enumerating all of the variables, later steps will deal with defining the variables more fully, anticipating problems in measuring and comparing the variables, and formulating a study design in which the measurements can be used for drawing casual inferences. In all studies, variables may serve any of three different uses. They may serve as dependent variables, independent variables, or extraneous variables.

Dependent variables

Also called: outcome variables
response variables
disease / effect variables

Independent variables

Also called: treatment variables
classification variables
casual variables
factor variables / risk factors

Extraneous variables

Also called: confounding variables,
biasing variables
noise variables
co-variables / effect modifiers

As an example, in a clinical trial to determine the efficacy of a investigational drug for preventing angina pectoris, the group to whic the patient was assigned (drug or placebo) would be the independe variable; whether the patient subsequently experienced angina pector (or the number of attacks) would be the dependent variable; and suc measurements as the cause of the anginal pain (coronary artery diseas versus valvular heart disease), the extent and severity of coronar arterial narrowing, and the number of previous myocardial infractior might be the extraneous variables of interest.

In identifying and describing the variables, the investigator must als anticipate the statistical analysis that will be applied at the end of th study.

Step 4: Define the Study Population

While determining the variables to be measured, the investigatc simultaneously defines the population to be studied. In clinical researcl the population is generally a group of patients found in the hospita admitted to a clinic, or simply identified in the population. In clinic research, outcomes are usually measured by using four types of scale to study the variables which are as follows:.

1. Categorical, dichotomous

   The variable has only 2 categories, such as died ("yes"/"no") c SEX ("male"/female").

2. Categorical, multicategoried, nominal

   The variable has more than 2 categories but they are namin (nominal) and intrinsically ordered such as COLOR ("white" "black"/"green"/"red") or SERVICE ("Medicine"/" Surgery "Obstetrics"/ "Pediatrics").

3. Categorical, multicategoried, ordered

   The variable has more than 2 categories and they are intrinsically ordered (ordinal), such as TITER ("non-reactive"/" 1:8"/" 1:32"/" 1:64") or

   AGE GROUPS ("18:34"/"35-49"/"50-74"/"75 and older").

4. Continuous (or interval)

   The variable is measured on a continuous scale, such as TEMPERATURE (in degrees F), or has a large number of discrete, ordered values, such as AGE (in years).

Once the target population has been described, the investigator must decide whether to enroll all members of the target universe into the study or to select a small sample of them. In most research, it is not feasible to study all members of the target population, and therefore samples of the target population are almost always studied. In selecting the sample, the investigator must weigh the alternative types of samples carefully to ensure that the final results can be generalized back to the target universe.

There are three basic kinds of sample. A statistical sample is a sample of the target universe selected by some process of random numbers (Described in detail in the chapter – Process of randomisation in clinical trial). . The advantage of a statistical sample is that the results obtained in the sample will be true of the target population within a range of error that can be calculated from the information in the sample.

The second type of sample is convenience sample. It generally comprises a group of patients from the target population, non randomly selected, who could be conveniently enrolled in the study and who seem to make up a reasonably representative sample of the target universe.

The third type of sample is referred to as chunk. When random sampling is not feasible and there are no convenient samples, an investigator may be forced to study a group of subjects similar to the target universe. When studying a chunk, the investigator must admit that the results might give a highly misleading picture of what is true in the target universe.

The 3 types of samples for making efficient estimates of parameters in a large target universe

1. **Statistical sample:** Selected by a system of random numbers. The results are true of the target universe with known error.

2. **Convenience sample:** Selected by a convenient, though, non-randoı method. The results are probably true of the target universe but tl error is not precisely known.
3. **Chunk:** Selected by a convenient, though non-random, method. The is no way to tell whether results are true of the target universe, and tl error is entirely unknown.

Step 5: Refine the Objectives into Written, Testable Hypothesis

Once the objectives have been written, the variables defined, and tl study population determined, it is necessary to refine the written objec into a set of written, testable hypothesis. This must be done before tl study design can be determined. Basically, a hypothesis is simply prediction of how the variables will be found to be associated in tl final statistical analysis. It differs from the written objective in that is stated in terms of the actual statistical associations that will be fou among the precisely defined variables formulated in Step 3. The w the hypothesis is expressed has important implications on how the stu is actually designed. This is why the hypothesis must be quantitative stated before the study design can be determined. (Further details Hypothesis testing have been discussed in Chapter on – "Clinic Biostatistics").

Step 6: Anticipate Error and Bias

Another set of issues that must be explored before determining tl study design is the possibility that the results of the study might rendered inaccurate or misleading by certain types of error or bi that will occur in the data when they are collected. All scientific studi are subject to error and bias, irrespective of how well they are design and carried out. Poorly designed studies, however, are usually mo seriously affected, and these distorting influences may invalidate tl results and render a study useless and unpublishable. To minimize tl occurrence of error or bias in a study, it is crucial to anticipate tl types of error or bias or allow the investigator to control them in tl statistical analysis.

There are basically four types of error or bias in research studie These are sampling error, selection bias, information bias, aı confounding factors.

1. **Sampling Error:** A distortion in the estimate of an association betwe two variables, resulting from chance variation in the selection of tl sample. It usually occurs when the study sample is too small.

2. **Selection Bias:** A distortion in the estimate of an association between two variables, resulting from the manner in which subjects were selected for the study sample.
3. **Information Bias:** A distortion in the estimate of an association between two variables, resulting from measurement error or misclassification of subjects.
4. **Confounding Factors:** A misleading (spurious) association that results when the casual effect of an independent variable on a dependent variable is inflated (or deflated) in a particular set of data by the presence of a third casual variable that is colinked with both the independent variable and the dependent variable.

Hence, before finalizing the study design, the investigator must think deeply about the types of error and bias that are likely to be encountered in the design and analysis. It is far preferable to think of them at this point, while something can be done, than to learn about them from the anonymous reviewer after the scientific manuscript has been rejected. (Further details of Bias and Error have been discussed in Chapter on – "Clinical Biostatistics").

Step 7: Develop the Study Design

It is often tempting to decide on the study design earlier in the planning process, but this should be avoided because the main function of the study design is to address the issues raised in the first six steps.

The Purpose of the Study

In developing the most appropriate design for a study, it is important to identify consciously the purpose of the study. Generally, the purpose of studies can be divided into descriptive, hypothesis-generating, and hypothesis-testing. In descriptive studies the purpose is simply to describe the frequency or magnitude of some event or disease. If the study's purpose is hypothesis-generating, one usually relies on a "quick and dirty" observational study, most often a cohort or case-control design. For the more definitive hypothesis-testing purpose, experimental studies (prospective clinical trials) or observational studies of a cohort or case-control study can be used.

Important Study Designs

Study designs can be divided into three main types: surveys, experiments and observational (epidemiological, quasi-experimental) studies.

Relationship between the purpose and Design of Research Studies

Purpose	General Design Type Most Often Used
Descriptive	Survey, usually utilizing a sample
Hypothesis-generating	Cross-sectional survey or "quick and dirty" cohort or case-control study
Hypothesis-testing	Experiment (clinical trial) if random assignment is feasible and ethical or Observational study, often a cohort nor case control design, employing extensive design features to avoid or control for bias, but more elaborate designs are sometimes used.

The experimental study designs provide a powerful means of testing causal hypothesis in circumstances where treatments or risk factors can be assigned to patients or experimental animals by the investigator.

In general, experimental study designs are easier to plan, conduct, and analyze and are less susceptible to error and bias than are observational studies. Observational studies, however, are often less expensive to perform then are experimental studies and may represent the only feasible alternative when experimental designs are not feasible or unethical. In general, a high degree of expertise in epidemiological methods, statistical techniques, and computing skills is required to perform observational studies well. This is because, when randomization is not possible, the investigator must go to much greater lengths to control for the types of error and bias that are predictably present in observational data. On the other hand, error and bias can also creep into experimental designs, making it desirable to obtain the assistance of a qualified statistician in designing and analyzing clinical studies.(For Further details refer to chapter on "Clinical Research Design".

Step 8: Estimate the Sample Size Needed

Determination of the optimal sample size for a study is a critical feature of the design. If too few subjects are chosen, the study is likely to fall prey to sampling error, ending in either a Type I error or a Type II error. On the other hand, selecting too many subjects may increase the cost of the study beyond the financial resources available. To avoid either extremes, the investigator must predict the optimal number and uses that in selecting the subjects for study.

In most instances, new investigators should consult a qualified statistician to assist in estimating the required sample size, since the exact method depends on the type of study design selected. In general, though, for each type of study design, there are only a small number of parameters that must be measured, estimated, or guessed to form the basis for estimating the optimal sample size. With experience, one can often arrive at a satisfactory sample size estimate by entering the required parameter estimates into the sample size program of one of the standard statistical packages for personal computers. Further details of Sample size calculation refer to Chapter on – "Clinical Biostatistics").

Step 9: Write A Research Protocol for all Collaborators to Review

It is crucial that the investigator writes a description of the study plan, before beginning work on the project. By writing the protocol, the investigator clarifies the exact intent and details of the proposed plan. In the absence of a written protocol, crucial details or concepts will be confused, vague, or frankly conflicting, and many of these problems will not be noticed. By writing down the protocol early in the study design process and passing it among the collaborating investigators, the basic objectives and the design can be agreed upon. Subsequently, as further steps are completed, the protocol can be extended to include the additional elements of the design. In this way, all collaborators can remain updated on the state of the design at any time.

Step 10: Plan the Data Collection Process

It is important to plan the data collection process in great detail to ensure the accuracy of the measurements of the independent variables, the dependent variables, and the extraneous variables. Forms used to record the data should also be developed and, if appropriate, tested for accuracy and clarity in a pilot study ahead of the main data collection process.

Step 11: Manage and Monitor Data Collection

Most importantly, the investigator must learn to distinguish between what is scientifically right and what is easy. Compromise often results in misclassification of subjects. If the misclassification is merely random, it reduces the chance of finding a significant difference. If it is systematic and related to the hypothesis under study, it constitutes a bias and will distort the study's results.

Particularly useful tools for planning and managing the data collection process are the PERT (program evaluation and review technique) charts and Gnatt charts. Both utilize charts showing the various tasks or phrases, on a time scale. PERT charts display the various tasks as "nodes" connected by lines between related tasks, whereas Gnatt charts display the tasks as rectangular boxes arranged on a time scale and extending from the beginning to the time of completion of the task.

As data collection progresses, the investigator must continue to monitor the accuracy of the data. This monitoring function may be as simple as periodically looking over the shoulder of a laboratory technician or as extensive as having certain percentage of the subjects studied independently by two or more of the data collectors as an in-line check on the reliability of the measurements.

Step 12: Manage the Database

The database is the quantitative record of the values of the dependent, independent, and extraneous variables measured on all of the subjects in the study. In small clinical studies involving only a few patients, the data can often be effectively managed solely on paper forms or by recording in notebooks. Larger clinical studies, however, require storing the data in a computer. Moderate size databases can often be managed by the investigator using one of the commercially available database management software packages. The management of large databases may require the assistance of a staff of trained database programmers. The larger the size of the database the more important it is to set up procedures for accounting for the data collection forms to avoid losing them, entering and editing the data to minimize data entry errors, and backing up the database to avoid losing it an archiving accident.

Step 13: Analyze the Results into Tables and Graphs

While new investigators often tend to become preoccupied with statistical tests of significance, the real heart of analyzing the data from a clinical research project is conceiving and producing the tables and graphs that most effectively test the hypothesis and describe the results. Ideally, the most important tables and graphs are conceived ahead of time in the design phase, but in most studies some amount of additional analysis is necessary to find the most enlightening way of displaying the results so that they will convey the findings most powerfully to the journal reader. For example, exploratory analysis is often essential for testing the normality of distributions of continuous variables, identifying extreme observations ("outliers") that would bias the calculation of means and perturb statistical tests, testing for confounding variables, and performing poststratified analyses to control for biasing factors. Much of this work depends on the content of the database and cannot be entirely anticipated ahead of time.

Almost always the analysis of the research database is done with one of the many computerized statistical packages. At first the investigator should work closely with a statistician to ensure that the analyses are done correctly and completely. After more experience, the investigator may assume progressively more of the analytic responsibilities to get more of his or her intuition into the analytic process. Even the most experienced researchers, however, continue consulting qualified statisticians on important aspects of their analyses and ask them to review draft manuscripts from a statistical perspective before submitting them to scientific journals. Since well-qualified statisticians often have preferred methods and approaches to analyses, it is wise to involve the statistician, who will assist with the analyses of the results, in the design phase of the project.

Step 14: Test the Hypothesis with the Correct Statistical Tests

Early in the design phase, it is important that the investigator anticipate the statistical test(s) to be applied to the final results. Even though the statistical process will ultimately prove more extensive, the basic statistical approach should be thought through early in the design. To choose the right statistical test in the majority of circumstances, the investigator must ask four simple questions. The answers to these four questions will determine the most appropriate statistical test in the majority of instances.

Firstly, "What are the variables being compared?" To answer this question, the investigator identifies the independent variable(s) and the outcome variable that are to be compared in the central hypothesis test at the end of the study.

Secondly, when the variables are identified, "What is the form of these variables?" The investigator identifies what type of variable each is. Is each a categorical, dichotomous variable; a categorical, multicategoried, nominal variable; a categorical, multicategoried, ordered variable; or a continuous (or interval) variable?

Thirdly, if the outcome variable is continuous, "Does it meet the normality assumption?" This question is answered by graphing the distribution of the outcome variable in the population to be studied. This can often be done by data obtained on this or a similar population from previous studies or from the literature or the investigator may wish to perform a pilot study to estimate the distribution.

Finally, "Are the comparison groups independent or paired?" In simple experiments, most cohort studies and some case-control studies, the observations on the study subjects are independent of one another, and the answer to the question is "no". In repeated measures of experimental designs and in matched case-control studies, the observations on the subjects are paired, in which case the answer to the question is "yes".

(Further details on Selection of statistical tests – refer the Chapter on – "Clinical Biostatistics").

Step 15: Generate Additional Tables and Graphs to Explain the Findings

After the main test is performed, the investigator or the consulting statistician will generally perform additional analyses to elaborate on the findings. For example, after finding an association between the independent and the dependent variables, it might be desirable to perform tests to demonstrate a dose-response relationship or to repeat the main analyses controlling for extraneous variables. These additional analyses should be anticipated in the design phase so that the variables will be appropriately coded and the sample sizes will be adequate for them.

Step 16: Publish the Results in a Scientific Journal

Finally, the end product of a successful study is the publication of an article in a scientific journal. Remember if a study is not published, it was never done. Successful writing for scientific journals is an art form that must be acquired through study and practice. It is different from other styles of writing.

One should always decide on a specific journal before starting to write a scientific article, consult the "INFORMATION For AUTHORS" page that appears in most journals, customize the article according to the instructions, and photocopy several articles from recent issues of the journal to serve as models.

Generally, a good scientific article presents no more than five or four main points that are key findings in the results section. These should be illustrated by tables and graphs in nearly final form before sitting down to write. The writing process goes much smoother if the tables and graphs are completely laid out and the results are written to describe and relate to them.

Clear writing requires an outline, and papers written without an outline are often disorganized. After the initial draft is made, it should be revised over and over again and passed among colleagues for critical review and revision.

Conclusion

Designing clinical research projects is a creative and rigorous process that requires attention to a large number of details, each of which can determine the success or failure of the project. For the planning and execution of projects to be successful, the investigator must be fully aware of the many aspects that must be planned so that all are addressed as early as possible in the research process. The steps outlined in this chapter provide a road map to give new investigators a framework and a checklist of issues to be addressed while planning their research projects. Proficiency in designing research requires study of more detailed references on these subjects and the guidance of experienced colleagues and mentors.

Suggested Reading

1. MUR Naidu, P Usha Rani. Designing Clinical Trials. Basic principles of Clinical research and methodology Ed SK Gupta. Jaypee Publisher 2007
2. Armitage P. Controversies and achievements in clinical trials. Control Clinical Trial 1981; 5: 243-51.
3. Boissel JP. Impact of randomised clinical trials on medical practises. Controlled Clin Trials 10; Suppl; 1989; 1208-348
4. Meinert CL. Clinical trials: Design, conduct and analysis. New York Oxford University Press 1986
5. Shapiro SH, Louis TA (Eds). Clinical trials: Issues and approaches. New York 1983; Marcel Dekkar.

CHAPTER 7

Clinical Research Design

G. Venkat Ramana, M.U.R. Naidu, P. Usha Rani

7.1 Introduction

The aim of this chapter is to provide a guide and reference to anyone concerned with the designing of clinical research. A good clinical research design will ensure that the study setup is achievable, the investigator will recruit clinical trial subjects /patients and the study will be completed within the target time lines. Failure of a trial design to achieve any of these steps may have direct consequence on the outcome of the clinical research. A failure of one part of the clinical research program may lead, at the least to a delay in reaching targets or can even go up to the cancellation of the program. Major consequences would be the cost, time, resources and most importantly loss of potentially useful drug. Hence the success of drug development process will greatly depend on scientific and ethical design of a clinical research.

The basic objective of a good clinical trial design would be:

- To minimize bias.
- Avoid ethical and practical issues associated with the use of investigational product/device.
- To ensure validity, reliability, generalizability and ability to replicate experimental results.

7.2 Elements of Trial Design

This section of the chapter describes the elements that make up a clinical trial and suggest how these should be considered when designing a trial and/or reviewing the design presented.

7.2.1 Subject / Patient Population and their Recruitment in to the Study

Recruitment of an adequate and appropriate number of study subjects is essential for the successful completion of a clinical research project. It is very important to define carefully the patient population from which the study participants will be selected. Proper selection will avoid bias and ethical issues that, derives from non defined, perhaps investigator specific patient population. It is unethical to study new chemical entity when more effective treatment options are already available for that particular aliment. However, the criteria of selection can be the treatment resistant condition where in there will be no other options for the management. In early Phase II trials the entry criteria may be limiting, and throughout the drug development programme, the criteria will be altered as the characteristics are discovered. Another classical example would be the volunteer selection in Phase I trials, where in there will be restricted age range, usually between 18 – 45 years. As further clinical trials are carried out in older, and possibly younger subjects, the treatment protocols may be extended to include these age groups. Usually by the end of Phase III studies the trial subjects should include patient population which is representative of the population who will receive the drug; so that the results have generality. Defining eligibility criteria will overcome bias of the study.

Influence of disease being studied is very important in avoiding variability which is inherent within any population or sub-population. In psychiatric clinical trials, simply enrolling any patient with symptoms of depressive disorder as per Diagnostic Statistical Manual-IV (DSM-IV) will introduce a wide range of variability.

Ideally, participants should be enrolled at a constant rate to maintain study power and minimize uneven or excessive workloads during the follow-up period. Participants who are likely to adhere to all aspects of the study protocol should be identified, if possible on pilot basis to improve the recruitment process.

7.2.2 Effect of Indication on Trial Design

Another important aspect to be considered while designing a trial would be the nature of the disease being treated versus. The duration of the trial and the timing of each patient's treatment. The nature of the disease also dictates the number of days of the study and timing of assessments.

Diseases under study may be broadly classified as acute and chronic illness. Acute indications of the drugs like infectious disorders or musculoskeletal injuries will definitely require short term designs and at this period the cure would be expected or at least some improvement seen. These types of diseases are also expected to have spontaneous remissions within certain duration. In contrast, those studies which are conducted for chronic illness will run from months to years. Trial design should be specific for the phase (acute, continues, maintenance and discontinues) of treatment under study. For example, in the case of schizophrenia and other psychiatric illness which are chronic and have relapsing course, these trials should not be of short duration to evaluate the efficacy for specific symptoms such as negative or cognitive dysfunction and should also address important questions regarding maintenance of overall clinical symptoms and long-term side effects such as tardive dyskinasia. The other important issue that has to be addressed in trials involving chronic illness would be concomitant medications that the patient might be already on. This may necessitate a wash out period. Any withdrawal of previous therapy is carefully monitored so that patient should not be destabilized. If a washout is used, trial design should incorporate various assessments and monitoring strategies so as to safeguard patient's safety.

In some cases particularly in chronic diseases the withdrawal of any current therapy can have better scientific outcome, but proper justification is vital before withdrawal. However, there can be several approaches in addressing these issues, one of the approaches is to recruit only patients who are recently diagnosed with the specific condition, this will mean that the number of patients available for the study will be reduced drastically. Increasing the sample size will address this problem. Another way to avoid withdrawal is to design a trial where the investigational drug is added to the current treatment. In such designs, disease under study is treated with highest possible dose of standard treatment to see the desired effect, still if the satisfactory effect either in terms of efficacy and safety is not achieved, then investigational drug can be added. In such studies there are possibilities of undesirable adverse events and drug interactions. Use of placebo in such situations can overcome this problem. Following a trial that proves efficacy and safety of the combination treatment, it might be appropriate to design trials that investigate the potential for decreasing the dose of standard medication without compromising the efficacy.

7.2.3 Randomization

Next important element and which is considered to be the heart of clinical trial design is *Randomization.* Detailed description about randomization is described elsewhere in this text book The terms 'random' and 'randomization' describe the premier mechanism currently in vogue for achieving group equivalence. These terms may connote different things to different people. The term 'random' means that every unit has equal chance of being selected in a sample. The term 'randomization' means the selection of subjects in a manner that controls the effects of extraneous variables in a manner determined by chance. Randomized clinical trials are considered to be gold standard for comparative medical studies, especially when they are free from missing data and non compliance.

7.2.4 Choice of Control

The goal of any successful design of a clinical trial is to minimize and attempt to eliminate most, if not all human bias. In this regard choosing a control group is a vital decision in designing a clinical trial. The choice of control affects the inference that can be drawn from the trial and the ethical acceptability of the trial. The basic principle of a controlled clinical trial is to assemble groups of people that are truly comparable as all aspects may influence outcome. By choosing the control group, it is possible to know what would have happened to patients or healthy volunteers if they have not received the test treatment or if they had received a different treatment that is known to be effective. In some situations, the course of illness is in fact predictable in a defined population and it may be possible to use a similar group of patients previously studied as a historical control. In most situations, however, a concurrent control group is needed because it is not possible to predict outcome with adequate accuracy or certainty.

Control groups in a clinical trial can be classified on the basis of two components

1. The type of treatment used.
2. The method of determining who will be in the control group.

The types of control treatment may be any of the following

(i) Placebo control
(ii) No treatment control
(iii) Dose response control
(iv) Active control.

Each type of control group is appropriate in some circumstance, but none is adequate in every situation. Following section describes briefly different types of controls.

(i) ***Placebo Control:*** In these studies subjects are randomly assigned to a test treatment or to an identical-appearing treatment that does not contain the pharmacologically active ingredient the treatments may be titrated to effect or tolerance, or may be given at one or more fixed doses. Such trials should be double-blinded. In some situations where more than one dose of the test treatment is required and in situations where the test formulation is a parenteral preparation and active comparator is an oral formulation, use of 'double dummy' techniques is appropriate. It is important to note that not every study that includes placebo is a placebo controlled study. For example, an active control study could use a placebo for each drug (double-dummy) to facilitate blinding; this is still active control trial, not a placebo-controlled trial.

One of the important concern using the placebo as a comparator is ethical issue. Use of placebo control may raise problems of ethics, acceptability, and feasibility, when a proven effective treatment is available for the condition being treated. Another important issue which requires the use placebo controlled trials is when efficacy testing is subjective.

(ii) ***No-treatment Control:*** In a no treatment trial, subjects are randomly assigned to test treatment or to no (i.e., absence of) study treatment. The principle difference between this design and a placebo controlled trial is that subject and investigators are not blind to treatment assignment. Because of the advantages of double-blind designs, this design is likely to be necessary and suitable only when it is difficult or impossible to conduct double-blind studies (e.g., treatments with easily recognized toxicity) and/or only when there is reasonable confidence that study end points are objective and the results of the trial are unlikely to be influenced by the variable factors.

(iii) ***Dose-response Control:*** In a randomized, fixed-dose, dose-response trial, subjects are randomized to one of several fixed-dose groups. Subjects may either be placed on their fixed dose initially or be raised to higher doses gradually, but the intended comparison is between the groups on their final dose. Dose-response trials are usually double-blind. They may include a placebo and/or active drug. In a concentration – controlled, treatment groups are titrated to several fixed-concentration windows, this type of trial is conceptually similar to a fixed-dose, dose response trials.

(iv) ***Active Control:*** In an active control (or positive control) trial, subjects are randomly assigned to the test treatment or to a active control treatment. Such trials are also usually double-blind, but this is not always possible; many oncology trials, for example, are considered difficult or impossible to blind because of different regimens, different routes of administration and different toxicities. The main objectives of the active control trials can be to show the efficacy of the new treatment by showing it is as good as a known effective treatment or superiority to active positive control. They may also be used with the primary objective of comparing the efficacy and/or safety of the two treatments.

7.2.5 Blinding

In a clinical trial, bias is one of the main concerns, it may be defined as systematic error or difference between the true value and that actually obtained due to all causes other than sampling variability. It can be caused by conscious factors, subconscious factors, or both. Bias can occur at several places in a clinical trial, from the initial design, operation during data analysis and interpretation. The general approach to the problem of bias is to keep the participant and investigator blinded, or masked to the identity of the assigned intervention.

Types of Blinding

Unblinded Trials

In an unblinided (also called as open label) trial, both the participant (trial subjects) and investigator know to which intervention the participant has been assigned. Unblinded studies are easy to conduct than other trials for two reasons (a) all other things being equal, it is simpler to execute than other studies (b) The trial is easier to design and carryout, and consequently less expensive, if blinding is not an issue. The main disadvantage of an unblinded trial is the possibility of bias. Participant reporting of symptoms and side effects and prescription of concomitant or compensatory treatment are all susceptible to bias.

Single-Blind Trial

In a Single-blind study, only investigators are aware of which intervention each participant is receiving. The advantages of this study are similar to those of the unblinded study. It is simpler to carry out than a double-blind design, and the knowledge of the intervention may help the investigator exercise their best judgment when caring for the participants. The disadvantages of the single-blind design are similar to those of unblinded design. The investigator avoids the problems of biased participant reporting, but they themselves can affect the administration of non-study therapy, data collection, data assessment. Both unblinded and single-blind trials are vulnerable to another source of potential bias introduced by investigators. This relates to the group differences in concomitant and compensatory treatment. For example in psychiatric clinical trials, since the assessment of response is completely subjective there is a possibility of introduction of investigator bias.

Double-blind Trials

In these studies neither the participants nor the investigators responsible know the identity of the intervention assignment. Such design is usually restricted to trials of drug efficacy, also called as double masked study. Participants may well see the treatment being given to participants in the other treatment group(s), and the appearance of the drug used in the study could give a clue to its identity. Differences in taste, smell, or mode of delivery may also influence efficacy, so these aspects should be identical for each treatment group. Even colour of medication has been shown to influence efficacy. Blinding is possible using the "double dummy" method. For example, if we want to compare two medicines, one presented as green tablets and one as pink capsules, we could also supply green placebo tablets and pink placebo capsules so that both groups of patients would take one green tablet and one pink capsule. In a double blind trial certain functions in open or single-blind studies could be accomplished by the investigators, must over taken by the others to maintain the blindness. A person (Study Pharmacist) other than investigators who sees the participants needs to be responsible for assigning the interventions to the participants. The credibility of the study will be increased in these designs particularly when the outcome measure is subjective.

These are more complex and difficult to carry out than other trials. One must ensure that the investigators remain blinded and that any data that conceivably might endanger blinded be kept away from them during the study. A data monitoring scheme must be set up and an emergency unblinding procedures must be established. These may increase the cost of the study.

Triple-Blind Trial

A triple-blind study is the extension of the double-study design, although the meaning of this may vary according to the exact study design. The most common meaning is that the subject, researcher and person administering the treatment (pharmacist) are blinded to what is being given. Alternately, it may mean that the patient, researcher and statistician are blinded. These additional precautions are often in place with the more commonly accepted term "double blind trials", and thus the term "triple-blinded" is infrequently used.'

Unblinding of Clinical Trials

Unblinding the trial or breaking the blind should be performed officially. There will be situations that require disclosures especially in long-term trials or studies. The study drug requires tapering the dosage, in emergency knowledge that a participant is or is not on active drug would indicate whether tapering is necessary. In case of emergencies withdrawing of the medication without breaking the blind will be used most commonly. If physician is different from the investigator, a third party can obtain the blinded information from the pharmacy or central data repository and relate the information to the treating physician. When unblinding does occur, the investigator should review and report the circumstances that led to it, knowledge of the kind of intervention, seldom influences emergency care of the participant to the IRB or IEC, such reviews will help reduce the frequency of the further unblinding.

A procedure should be put in place to break the blind quickly for any individual/ participant at any time should be in the best interest of the participant. Systems include having labels on the file in the hospital pharmacy or other accessible locations, or having "on call" 24 hours a day someone who can decode the assignment. To avoid needless breaking of the code, someone other than investigator could hold a list that reveals the identity of each drug code.

Alternatively, each study medication bottle label might have a sealed tear-off portion that would be filled in the pharmacy or with the participants records. In case of emergency the seal could be torn and the drug identity revealed. Care should be taken to ensure that the sealed portion is of appropriate colour and thickness to prevent reading through it.

In summary, double-blind trials require careful planning and constant monitoring to ensure that the blind is maintained and the participant safety is not jeopardized.

7.3 Types of Clinical Trial Design

Prospective and retrospective clinical trials are the major type of clinical trial designs. Most popularly used clinical designs with great scientific validity are prospective designs. In these studies all patients or healthy volunteers will undergo specific intervention and will be followed for a specified duration of time to the see response. The main advantage of these designs would be to minimize the interference of confounding factors while interpreting a clinical trial.

Fig. 7.1 describes different types of clinical trials which are commonly used in practice.

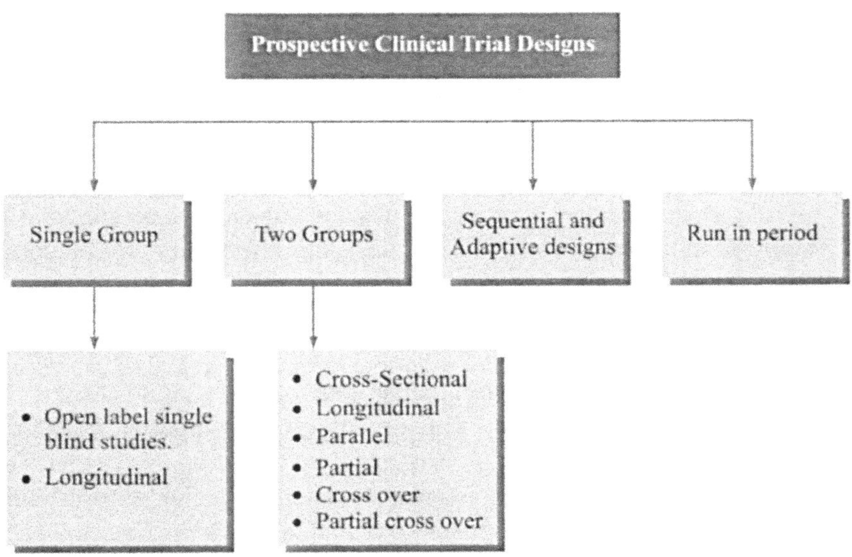

Note: Different designs that the mentioned above are only most commonly used designs.

Fig. 7.1 Clinical Trial Designs

The different types of clinical trial designs are:

(a) Single group trial design
(b) Two group trials longitudinal design
(c) Cross sectional trial design
(d) Parallel and cross over trial design
(e) Sequential trial design
(f) Run-in period trial design
(g) Adaptive trial design

(a) ***Single Group Trial Design:*** Open label single blind trials are the commonest examples for these types of designs. In these trials there will not be control group. Usually these are open label trials. Most of the times these trials are pilot studies usually to establish therapeutic activities of drugs.

(b) ***Two Group Trials Longitudinal Design:*** Longitudinal trials are usually long-term clinical trials conducted for several months or longer. These trials are most commonly active controlled trials and placebos are less used. However, 'no treatment' groups are some times included in these trials. The basic objective of longitudinal trial is to compare rates of change of a continuous longitudinal response between two treatment arms. Such comparisons are common in studies assessing the change in disease marker, such as lung function (FEV_1) in respiratory studies, blood pressure in cardiovascular studies, and viral load in HIV (Human immunodeficiency virus) studies.

In psychopharmacology trials a frequently employed design for evaluation of long-term treatment efficacy is a placebo-substitution paradigm. This study design selects patients who are acute responders and randomly assigns them either to continue study medication or to switch to placebo, in a double-blind manner. This design is adaptable to evaluating prevention of relapse (same episode) or recurrence (new episode). Most studies of this type recruit patients during acute phase or episode of illness, and require that patients attain adequate control of symptoms (response) before randomization to long-term treatment with drug or placebo. This type of enrichment design has inherent limitations regarding generalizibility of results to an unselected patient population. Attrition from the acute treatment phase in these studies may reach 50% or more. This can produce patient samples in the long-term treatment phase of the study that are unrepresentative of the original intake criteria and end up inadvertently favoring one treatment over another (Fig. 7.2).

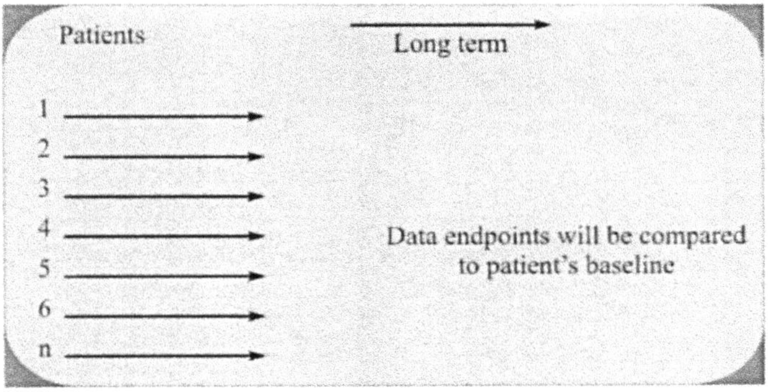

Fig. 7.2 Longitudinal Trial Design

(c) ***Cross-sectional Trial Design:*** These trials last for short term; where in a cross section of the patient population of interest is evaluated for a period up to 10 weeks. Longer cross sectional trials are also possible, but are less common. Patients are usually placed into one of two or more treatment groups and data obtained from each group are compared. For example, in understanding the efficacy of antidepressants pertaining to presenting somatic symptoms profile, we can follow all patients who are prescribed with different antidepressant without intervention. At the end of specified time we can evaluate only drugs which are commonly prescribed and then compare between them (Fig. 7.3 A and **B**).

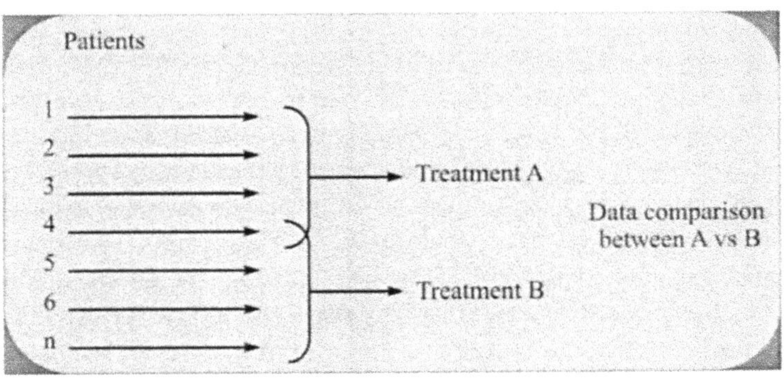

Fig. 7.3 A Cross-Sectional Trial

Example: Cross-sectional Study of Antidepressant prescribing patterns and response.

(d) ***Parallel and Cross Over Designs:*** In the parallel design, subjects receive one treatment for the duration of the trial. Each group generally receives either trial medicine or placebo or standard drug, or one of two doses of the same trial medicines. More specifically, a common feature of the parallel group design is to incorporate a baseline measure and a final

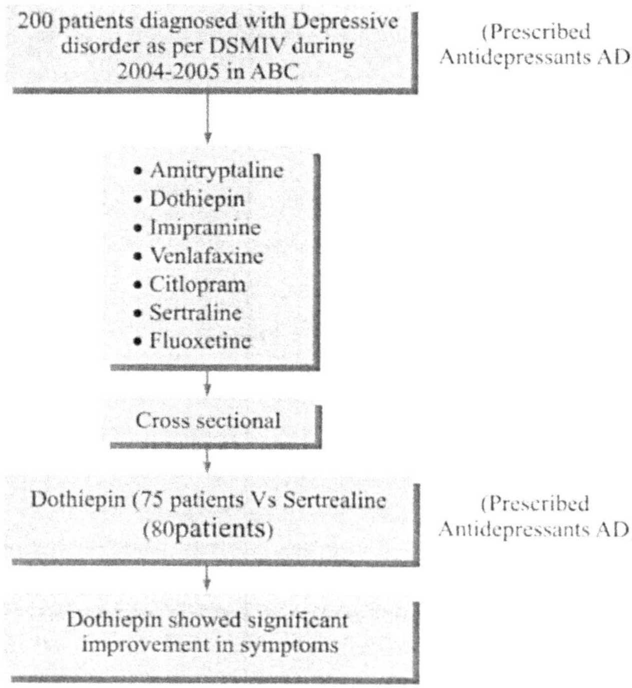

outcome measure. The final outcome variable in the parallel design may be defined in numerous ways depending upon the hypothesis of interest, e.g. change from baseline or percent change from baseline. The parallel design is applicable to most experimental situations. In contrast, the cross-over design allows subjects to receive two or more treatments during the course of the trial. For instance, in the standard 2 x 2 cross-over trial, subjects may initially be randomized to an investigational treatment or placebo/standard treatment and then at some predetermined point they will switch over to other treatment group. Cross-over design may or may not incorporate base line assessments. These types of designs are most important in exploratory clinical trials like Phase II to determine whether or not to proceed to a Phase III trial. In this design, each patient acts as their own control and in the clinical treatment of a patient this may be helpful in identifying which treatment is best suited for that patient.

Factors Influencing Trial Designs:

Factors that influence the selection of type of design are as follows:

- Patient safety.
- Disease type.
- Resources.
- Outcome of interest.

- Patient recruitment.
- Equivalence versus superiority.
- Logistical considerations.
- Washout period and carryover issues.
- Anticipated differential dropout rates.

For a given disease and drug, cross-over design may be used in an early phase of the investigation and a parallel design may be used in a later phase of the investigation.

In general the choice of a 2 × 2 cross-over design to a two-arm parallel groups design in superiority trials comes down to the perception by clinicians of greater efficacy, i.e. it is assumed that fewer subjects are needed to carry out a test of treatment mean differences or other measures of centrality for the cross-over design relative to the parallel groups design. The great importance of cross-over design over parallel groups trials is, that it can reduce at least twice as many as subjects required for a parallel group studies.

Another key factor that favors more to the 2 x 2 cross over design is that all subjects have the opportunity to receive both active/placebo/ standard drugs. This may offer inducement to participate because randomization to placebo is temporary.

Cross-over design requirements

- The patient must have stable disease during both treatment periods. This is possible for chronic diseases conditions like hypertension, insomnia, migraine or epilepsy etc.
- Similar baseline condition must be present at the start of each of the two treatment periods.
- Must not have any carryover effect after either treatment i.e., the disease manifestations should revert to the same baseline and that the effect of the treatment should disappear when either treatment is stopped.
- Determine whether the change from one treatment period to another will be time dependent (e.g., after 5 weeks on treatment A all patients will be switched to B treatment).
- Informed consent should include detail description cross-over design and switching of treatment A to B.

Distinguishing characters between Parallel and cross-over designs are presented in Table below.

Parallel Design Vs Cross-over Design

Parallel Design	Cross-Over design
Practically simple to perform	Practically difficult to perform.
Investigational product is given only to patients in one treatment arm	All patients have equal chance of getting investigational product.
Placebo is randomized to only one arm	Placebo is randomized to both the arms.
Interventions are possible in acute and chronic diseases	Interventions are not possible in acute illness.
Enrolling patients with period effects are possible.	Only patients with no period effects are possible
Enrolling patients with physical and psychological carry over effect is possible	Enrolling patients with physical and psychological carry over effect is not possible
Can be possible for drugs with longer half-lives. e.g., Bioequivalence study	Cannot be possible for drugs with longer half-lives. e.g., Bioequivalence study
More number of patients are required	Comparatively less number of patients are required.
Comparatively low scientific validity	High scientific validity.

The figure given below describe different parallel and cross-over designs.

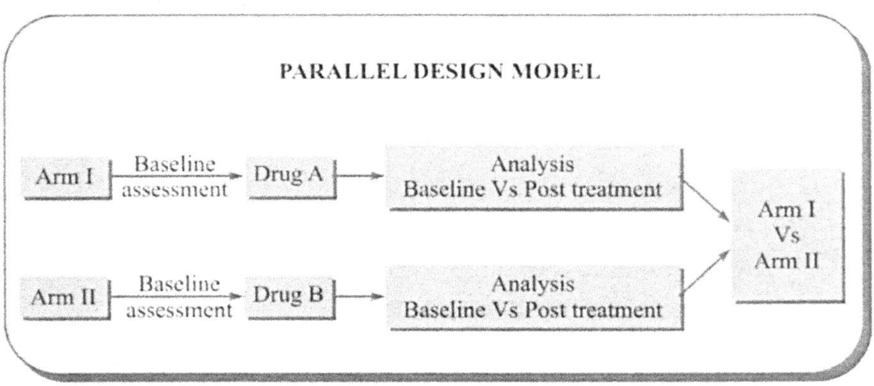

Cross-over Design Model

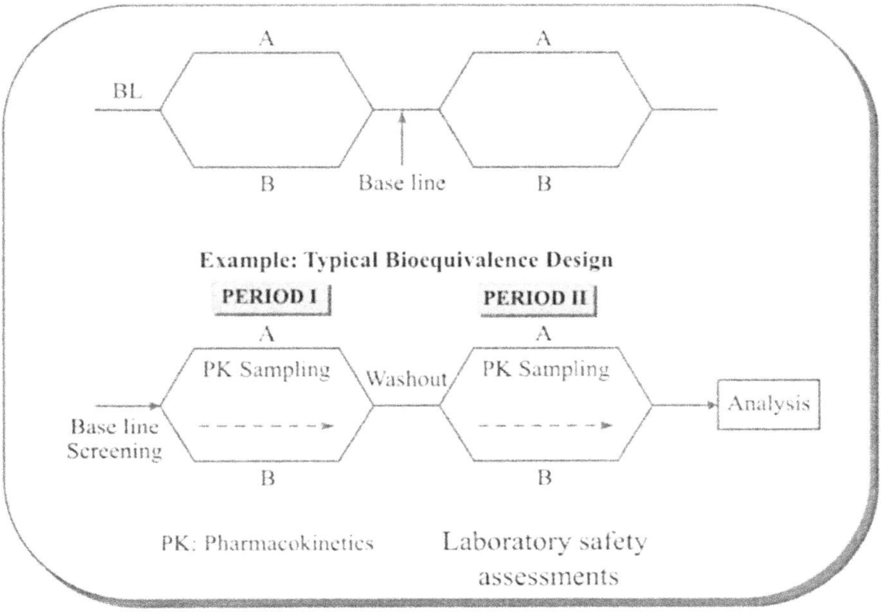

Presented figure is the selected schematic representation of a typical cross-over study that is commonly practiced. Example presented is a typical presentation of a bioequivalence study that has reasonably less half life. After baseline screening, subjects will be randomized to receive drug 'A' or 'B' in Period I and subsequently after 5 half-lives subjects crossed in Period II.

(e) ***Sequential Trial Design:*** Sequential designs are assigned to receive one of two treatments. Sequential balancing aims to achieve an equal number of subjects on each treatment in each class by considering all the factors in consecutive order. This design first tries to achieve balance on the first factor, then when it is successful, tries to achieve balance on the second factor and so on. This design also allows a series of interim analysis of the emerging data to be conducted, and specifies the circumstances under which the trial will stop or continue at each one. Such designs are likely to require fewer patients than traditional designs of equal power, and in particular can avoid continuation when one treatment is already evidently inferior to the other.

The major requirements for using this design are

- Rapid evaluability of patients.
- Simple trial design.
- Clear endpoint
- Limited hypothesis to be tested.

Major limitation of this design is that it is not suitable for most clinical trials. This design requires number of characteristics for this design to be appropriate, including simple trial design, limited scope of hypothesis and rapid evaluability of effects.

(f) ***Run in Period Trial Design:*** One of the important confounding factors in clinical trial is placebo responders and noncompliers. This becomes very important while conducting *exploratory* clinical trials (Phase II) based on which data *confirmatory* Phase III trials are generally decided. The goal of these studies is to evaluate the true activity of the investigational product. A run-in phase is a period prior to randomization during which potential participants who have met all entry criteria for randomized clinical trial are assigned the same regimen either the control (possibly placebo) or the experimental (active) treatment. An active run allows the dosage to be optimized prior to randomization, where as placebo run-in allows for both a longer drug-free interval and a more careful evaluation of patients before exposure to an experimental treatment. For these reasons run-in trials have become quite common in conducting clinical trials

Run in period Trial design

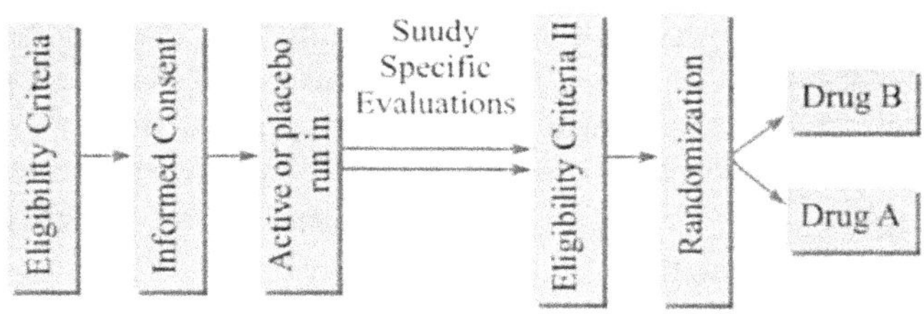

The above figure is a typical presentation of a run in period trial. In this design following Eligibility criteria I and informed consent subjects will be administered with either placebo or active/standard medication. Following study specific procedures for specified duration of time, subjects will be randomized to receive study medications.

It has been suggested that run in period may be particularly beneficial in primary prevention trials in which adherence is likely to be lower than in therapeutic trials, and in trials with infrequent contact between study staff and participants. Typically chemoprotective trials are lengthy and participants are not required to make frequent followup visits, Trial patients often tolerate minor yet noticeable side effects and committed to taking experimental drug with no guarantee of preventive effect. Therefore, chemoprevention trials are using run-in periods for screening. One such design is shown below.

Example of Run-in Period Trial Design

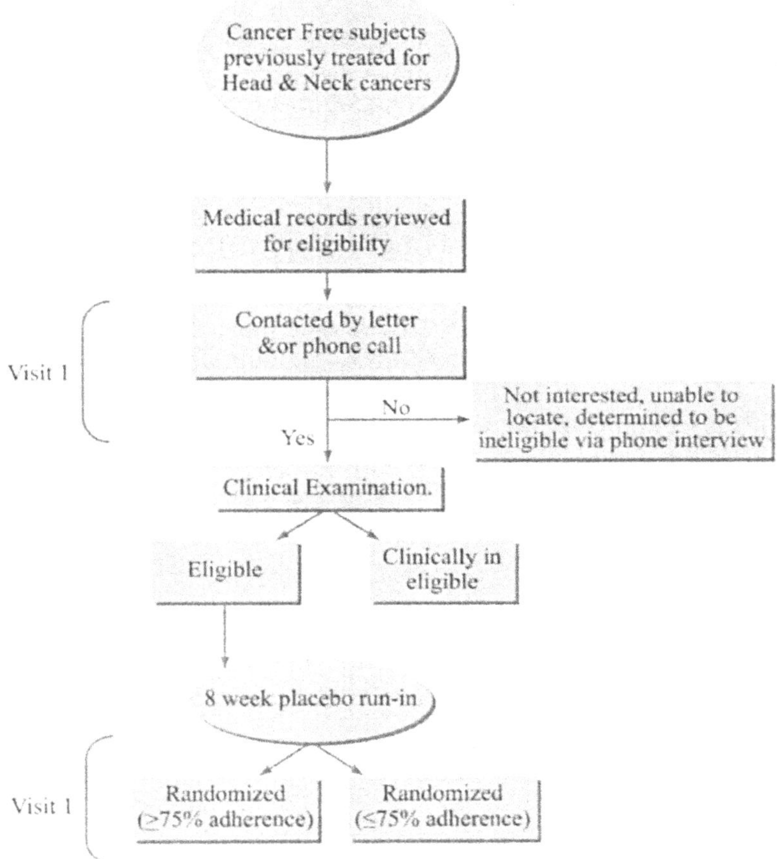

(g) ***Adaptive Trial Design:*** Conducting clinical trial is very expensive and time consuming. Failure of adequate response of the drug under trial is great loss to the sponsor. But at the same time sponsor and regulators do not expect false positive results. Careful conduct and design of the trial is most important to avoid any type of descripencies. *Adaptive design* is a flexible design for concuting clinical trials. It will allow mid-trial design modifications, which are based on interim information from inside or outside the trail while meeting regulatory requirements for the control of the Type I error (in detail described elese where in the chapter) rate. The basic principle is to combine stage standardization test statistics such as p-values or z-scores in a pre-specified way. Most statistical methods for clinical trials are based on the hypothesis testing, in which long-run probabilities for 'false positive' and 'false negative' test decisions and are controlled by fixing significance levels and sample size based on power calculations.

In adaptive randomization procedures the treatment allocation of the next patient is will depend on the outcome of the previous patients. Adaptive randomization in general sense permit various types of design modification during the course of the of the trial that are based on the observed interim results without increasing the false positive error rate. It is important for the protocol designer to include the combination rules well in advance in the protocol.

In principle, adaptive designs have common feature to group sequence trials in which the null hypothesis is tested in a sequence of interim analysis. Adaptive designs differ form group sequential designs by the possibility of redesigning the trial based on the full knowledge from interim analysis.

Conducting clinical research is undoubtedly very expensive and cumbersome. Scientific and ethical designing of the clinical trials will not only eliminate bias and other confounding factors but also helps in improving the outcome. Each trial design has its own advantages and disadvantages, thus selection of the specific design purely depends on the type of the study and expected outcome.

Suggested Reading

1. Shein Chung Chow. Design and analysis of clinical trials. (Concepts and methodologies). Wiley Publication.
2. Stephen B Hulley. Designing Clinical research. Williams and Wilkins publication.
3. John I Gallin. Principles and practice of clinical research. Elsevier publication.

CHAPTER 8

Clinical Research Protocol

MUR Naidu and P Usha Rani

Protocol **is** defined "As document that describes the objective(s), design, methodology, statistical considerations and organization of a clinical research/trial". The protocol is the detailed plan of the study.

The written protocol

- forces investigators to clarify their thoughts and to think about all aspects of study;
- is a necessary guide if a team (not a single investigator) is working on research;
- is essential if the study involves research on human subjects or is on experimental animals, in order to get the institution's ethical approval;
- is an essential component of a research proposal submitted for funding.

A well-thought out and well-written protocol can be judged by three main criteria.

- Is it adequate to answer the research question(s), and achieve the study objective?
- Is it feasible in the particular set-up for the study?

- Does it provide enough detail that can allow another investigator to do the study and arrive at comparable conclusions?

What is the Purpose?

Instructs study team on exactly what to do Explains the study to the outside world Provides a reviewable document for necessary approvals Many of the methodological aspects of designing a research study and writing a protocol can benefit from the advice of a statistician. Such advice should be sought at an early stage.

8.1 Title Page

8.1.1 Title

It is useful to specify both a full title and short title

- The full title should include summary study design, medicinal product(s), nature of the treatment, comparators and/or any placebo, indication, patient population and setting.
- The short title is a summary of this
- The titles specified must be consistent across all documents relevant to the trial

8.1.2 Names (Titles), Roles and Contact Details of

- Authors, investigators, experts and advisors involved in the trial
- Sponsor & monitor – as agreed with Chief / principal investigator's employer and the host Trust ?Trial site(s), clinical laboratory(s), technical departments and institutions involved in the study

8.1.3 Protocol Details

- Version number
- final / draft
- Date

8.2 Signature Page

Signatures of all healthcare professionals involved in the trial

8.3 Contents Page

8.4 List of Abbreviations and Definitions

8.5 Summary

1 or 2 page summary including:

- Aim and rationale for the trial
- Summary of trial disorder / interventions / measures
- Primary & secondary objectives
- Brief description of methods

8.6 Background

The detail given in this section should be backed up by a full literature review and should make reference to relevant papers, previous clinical experience and pilot work.

This section should include:

- A clear explanation of the main research question i.e. the hypothesis to be tested
- Detailed justification for the trial including :
    - explanation of why the study is appropriate, potential benefits to patients/ health service, relevance to current policies and priorities.
    - description of the indication, its diagnosis, incidence, current treatments and their limitations
    - description of the treatment under investigation including reference to any previous evidence of its usefulness
    - a statement of what would be a worthwhile improvement in study outcomes and
    - what evidence there is that the treatment under investigation may achieve this.

8.7 Trial Objectives and Purpose

- Purpose of research (e.g student project, commercial / non commercial trial, licensing)
- clearly define and distinguish primary and secondary objectives (including examination of effects for defined subgroups of patients)

8.8 Study Design

- statement of the primary and secondary endpoints / outcomes (including at what point in the trial these will be measured)
- clear description and justification of the type of design (e.g parallel group / crossover,

- sequential, cluster randomised and equivalence) if crossover design, include information about possible carry over effects, detail of orderings, washout (/in) periods etc
- Phase of the trial (e.g. phase I / II / III / pilot study)
- summary of treatments being compared with reasons for choice of comparison group (e.g active control / placebo)
- schematic diagram(s) of the trial design, procedures, stages and data collection
- description and justification of the duration of treatment, subject participation and trial follow-up

8.9 Subject Selection

Include detail of:

- Source of subjects (where they come from and why this group is appropriate)
- Number of centres involved
- Subject inclusion and exclusion criteria (with justification if necessary – for example consider contra-indications to trial treatments, incompatible concurrent treatments, recent involvement in other research)
- Expected no of eligible participants available per year and proportion of these expected to agree to the trial

8.10 Subject Recruitment

Details of recruitment process including

- method of recruitment (eg via adverts, clinics)
- payment of participants
- details of procedures, tests, screenings carried out to assess trial suitability
- Provision of patient information sheet (include as appendix)
- gaining patient consent (how consent will be obtained, who will gain consent, whether a witness will be present, how long the subject will have to decide, the arrangements for non English speakers and special groups (e.g. mentally ill, children, those suffering from dementia.)
- detail of enrolment procedure

8.11 Trial Interventions

This refers to the treatment under investigation and any active control treatment. Detail in this section may be referenced to other documents, such as the Investigators Brochure.

8.11.1 General Information

- full name, generic name (if appropriate).
- licence information - (as appropriate)
- the Summary of Product Characteristics or the Data Sheet for Licensed Medicinal Products.
- summary of known and potential risks and benefits to human subjects

8.11.2 Use during the Trial

- description and justification for the proposed route of administration, dosage, and treatment period
- detail of who will be administering the product (e.g. patient, nurse, doctor, carer)
- is the treatment invasive / does it involve radioactive substances?
- description of dosage form, packaging and labelling of products
- description of dispensing records, accountability and disposal procedures during the trial
- details of who will supply the products
- other detail including shelf life, arrangements for storage etc
- arrangements for continuation of treatment for study patients after the end of the trial
- Other medications permitted during the trial - include rescue medication (could be standardised for the purposes of the trial). Important also to consider possible interactions or effects that could confound results / conclusions

8.12 Randomisations

Including detail and justification for each of the following :

- patient / cluster randomised design (randomising individuals or groups (e.g. general practices, wards)
- type of randomisation to be used - simple, block, stratified, minimisation
  - if stratified include definition of stratification variables
  - if blocked define block sizes and whether these will vary.
- use of equal or unequal allocation between treatment arms
- information regarding how randomisation will be implemented (including who, where, how)
- approach to be used to conceal allocation (e.g. sealed envelopes, telephone central allocation office, computerised randomisation etc)

8.13 Blinding

8.13.1 Blinding

Detail and justification for:

- measurements to be blinded
- level of blinding to be used – e.g. blinding of participants / investigators / assessors (i.e. double blind, single blind, open)
- how blinding will be implemented (e.g. through use of identical placebo)
- Other measures taken to minimise / avoid bias

8.14 Data

8.14.1 Data Collection

- provide a detailed list of all data (outcome variables, explanatory variables etc) to be collected, with each description including :
  - source of the data (e.g. patient questionnaires, patient notes, electronic data, procedure)
  - time point for collection (baseline, during treatment, at followup point)
  - who will collect the data
  - why the data is being collected (e.g. baseline comparison data, main outcome, important prognostic / explanatory variable)
  - whether the data is from a standardised tool (e.g. McGill pain score) / involves a procedure (in which case full details should be supplied). If a non standard tool is to be used, detail on reliability and validity should be given.
  - what form the data will take (e.g binary, continuous (numeric), time to event)
- useful to include table / diagram describing schedule for data collection.
- describe methods used to maximise completeness of data (e.g. telephoning patients who have not returned postal questionnaires)
- include data collection forms as appendices

8.14.2 Data Handing and Record Keeping

- Describe procedures for data collection and recording (software to be used, location of the data etc)
- Detail methods implemented to ensure validity and quality of data (e.g double entry, cross validation etc)
- Security / storage of data
- Records retention – duration and location
- Statement of who is responsible for data collection, recording and quality

8.15 Statistical Considerations

8.15.1 Statistical Analysis

- Detail of the variables to be used to assess baseline comparability of the randomized groups and how these will be reported (e.g. means, standard deviations, medians, proportions)
- Detailed plans for statistical analyses of primary and secondary outcomes including:
  - summary measures to be reported
  - method of analysis (justified with consideration of assumptions of the method, structure of the data (e.g. unpaired, paired, hierarchical) etc)
  - plans for handling missing data, non compliers and withdrawals in analysis
  - plans for predefined subgroup analyses
  - Statement regarding use of intention to treat (ITT) analysis
- Detail of approach for interim analyses and criteria for early termination of the trial
- Detail of any non statistical methods that might be used (e.g qualitative methods)
- Statement of who will carry out analyses and at what point

8.15.2 Sample Size Calculation

- Details of the precision or power calculation used to estimate the required sample size (for analysis of the primary outcome), including : estimates used (e.g. size of the clinically important effect to be detected, drop out / non compliance rates) assumptions made (e.g. assumptions of Normality) relevant justification (i.e. appropriate references or clinical arguments) allowance for planned subgroup analyses chosen levels of significance and power methods / formula / software used
- An estimate of the recruitment period for the trial (calculated based on the expected number of eligible and recruited participants available per year) with justification that the required sample size will be attainable in practice.

8.16 Compliance and Withdrawal

8.16.1 Subject Compliance

- Procedures for monitoring (e.g. watching subject swallow pills and checking their mouths afterwards)
- Recording of patient compliance information (what will be recorded, when and where)
- Detail of follow-up of non compliant subjects

8.16.2 Withdrawal / Dropout of Subjects

- Describe under what circumstances and how subjects will be withdrawn from the trial
- Give details of documentation to be completed on subject withdrawal (including recording reasons for withdrawal and any follow-up information collected)
- Whether and how subjects would be replaced

8.17 Interim Analysis and Data Monitoring

8.17.1 Stopping / Discontinuation Rules and Breaking of Randomisation Code

- Define completion and premature discontinuation of the trial
- Describe procedure regarding decisions on discontinuation of the trial (e.g interim analyses, role of data monitoring committee)
- State documentation to be completed if part / all of the trial is discontinued
- Describe circumstances under which the randomisation codes may need to be broken and the procedure for this.

8.17.2 Monitoring, Quality Control and Assurance

- Use and role of monitors eg data monitoring groups and steering groups and arrangements for monitoring / auditing conduct of the research
- Assurance on good clinical practice and adherence to research governance guidelines
- Detail of any other steps taken to ensure quality of research

8.17.3 Assessment of Safety of Pharmocovigilance

- Definition of serious adverse events for the trial which are expected e.g. hospitalization in terminally ill patients.
- Statements about which serious expected adverse events will not be reported.
- A statement about how non serious adverse events will be recorded and reported.
- Details of the procedures that will be followed in the event of adverse events in the trial
    - who has what responsibility
- Methods and timing for assessing, recording and analysing safety parameters (e.g interim analyses)
- The type and duration of follow up for subjects after adverse events

8.18 Ethical Considerations

Description of ethical issues for the trial. For example consider:

- Approvals from relevant groups (e.g.IEC/IRB. DCGI. ICMR etc)
- Informed consent (append information sheet and informed consent form)
- Allowances for special groups (e.g non English speakers, children, mentally ill)
- Patient withdrawal / discontinuation
- Trial monitoring

8.19 Financing and Insurance

- Finance and insurance details (if not addressed in separate agreement)
- Cover for non negligent and negligent harm

8.20 Reporting and Publication

Detail of publication policy (e.g. Following the study, will there be access to raw data and right to publication freely by all investigators in the study?, what publications / conference presentations will be planned)

Tables, Figures, References

Appendices

Including (where relevant):

- Patient information sheet
- Patient consent form
- Data collection forms and validation information
- Summary of product characteristics
- Investigators brochure

Suggested Reading

1. Altman DG, Schulz KF, Moher D, Egger M, Davidoff, Elbourne D, Gøtzsche PC, Lang T for the CONSORT Group. The revised CONSORT statement for reporting randomized trials: explanation and elaboration. *Annals of Internal Medicine* 2001; 134: 663-694.
2. ICH Harmonised Tripartite Guidelines for Good Clinical Practice
3. Friedman LM, Furberg CD, DeMets DL. *Fundamentals of Clinical Trials*. 3rd edition. New York: Springer; 1998.
4. Jadad A. *Randomised Controlled Trials*. London: BMJ Books, 1998.
5. Pocock S. *Clinical trials: a practical approach*. Chichester: Wiley, 1983.
6. Norman M. Goldfarb. Protocol Evaluation Questions. Journal of Clinical Research Best Practise. Vol. 5, No. 4, April 2009

APPENDIX

Protocol Evaluation Questions

Study Design and Objectives

1. Is the protocol title clear and accurate?
2. Is this the final version of the protocol?
3. Does the protocol reflect competent preparation?
4. Are the objectives, rationale and activities of the study clearly stated and sensible?
5. Is the study justified by its scientific objectives?
6. Is this study similar to previous studies we have conducted, and what were the results of those studies?
7. What phase is the study?
8. Are prohibited activities required before informed consent is obtained?
9. Can all activities scheduled for each visit be completed in that visit?
10. What are the safety and efficacy endpoints for the study, and how are they measured?

Treatment

11. What is the study drug?
12. What is the comparator?
13. What is the ratio of subjects per arm?
14. How does study treatment differ from standard of care?
15. Is there a problematic washout period?
16. Are there technically difficult treatments, procedures or assessments?

Study Schedule

17. Is the schedule reasonable and practical?
18. When will we initiate the study and how does that date compare to other sites?
19. What is the enrollment period?
20. What is the follow-up period?
21. Will visits be performed outside normal business hours or create logistical challenges?
22. How often will the study be monitored and how will monitoring be conducted?
23. Are there any interim milestones, e.g., for data submission?

Subject Protection

24. What are the benefits, risks and burdens for the subjects?
25. Is the use of a placebo, if any, acceptable?

26. Does the study raise any standard-of-care issues?
27. Is the safety profile of the study drug acceptable?
28. Are there any other subject protection or ethical issues?
29. Will the IRB have issues?

Subject Recruitment and Retention?

30. Are the enrollment objectives – numbers and time period – realistic?
31. Do we have access to the study population?
32. Where and how will we find potential subjects?
33. Will they be interested in participating?
34. Is the study duration very long?
35. Are the visit schedule and times acceptable for subjects and practical for study personnel?
36. How much flexibility is there in visit dates?
37. Are there frequent, inconvenient or unpleasant procedures?
38. Is the dosing schedule inconvenient or complex?
39. Are subject diaries required?
40. Are the eligibility criteria too strict?
41. Are the eligibility criteria clear and unambiguous?
42. Are the screening procedures practical, limits and yields reasonable, and payments acceptable?
43. Are there seasonal issues?
44. Do we, or will we, have other studies competing for the same subjects?
45. Are other sites in the local area participating in the study?
46. Is enrollment competitive?
47. Are subjects from populations with extra recruiting and informed consent requirements?
48. When do subjects become evaluable?
49. Are subjects likely to stay in the study for its duration?
50. What subject recruiting support will the sponsor provide?
51. What costs will subjects incur?
52. What subject compensation makes sense?
53. After completion of the study, how and when can subjects get access to the study drug?

Adverse Events

54. How many AEs and SAEs should we expect?
55. What types of adverse events are recorded or reported?
56. What are the known side effects of the study drug?

57. What are the implications of pre-existing medical conditions for classifying adverse events?
58. How is the randomization code broken?
59. Who pays for treating subject injuries?

Personnel

60. Do we have qualified personnel available to conduct the study?
61. Does the study require above-normal time to conduct, e.g., visit lengths or CRF pages?
62. Is additional staff training required?
63. Are there unusual pharmacy, dispensing or accountability requirements?
64. Is specialist expertise required, e.g., for assessments?
65. Can the principal investigator delegate activities normally?
66. Are there any limitations on subinvestigators or satellite sites?

Facilities and Equipment

67. Do we have access to required facilities and equipment?
68. Are the laboratories local or central?
69. Are the arrangements for shipping lab specimens practical?
70. Are laboratory tests required when the laboratory is closed?
71. Do laboratory and other facilities have adequate capacity?
72. Do we have adequate space and equipment to store study materials and records?

Other

73. Do we have previous experience with or references for the sponsor?
74. Do we have previous experience with or references for the CRO?
75. Do we have ready access to sponsor/CRO personnel?
76. Does the study budget support our financial and other objectives, such as patient care, education, physician relations, and marketing?
77. Is the study sponsor financially sound?
78. Are there third-party pay or reimbursement issues?
79. Does the clinical trial agreement include problematic terms?
80. What study activities require pre-approval by the sponsor (e.g., unscheduled visits, extra lab tests, subject injury treatment, and advertising)?
81. Will EDC/eCRF be used?
82. Are there unusual requirements as to how study activities are performed?
83. Are there unusual reporting or other administrative requirements?

CHAPTER 9

Designing Case Report Forms

M.U.R. Naidu and P. Usha Rani

9.1 Introduction

In a clinical research, data from each participant should be recorded on a form specially prepared for this purpose known as case record form or case report form. One should take care to include all the information necessary to answer the questions asked in the protocol. Case record / report form (CRF) should be set out clearly and neatly. It should be simple to use and check and, if possible, should facilitate automatic data processing. As per ICH-GCP 1.11, a case report form is a printed, optical or electronic document designed to record all the protocol required information to be reported to the sponsor on each trial subject.

Case report form is a data collection tool used in clinical trials to support investigators and coordinators in capturing all protocol-required information. A well-designed CRF facilitates data collection and entry, and directly benefits other facets of data management and statistical analysis.

9.2 Uses of CRF

Case report form is a paper or electronic document designed to record all the information for an individual study subject required by the Study protocol.

The Case report form is used for several purposes:

- to ensure data collection in accordance with the Study protocol;
- to ensure fulfilling of the regulatory authorities' requirements for data collection;
- to facilitate the effective, comprehensive data processing and analysis, results reporting, and to promote the safety data sharing between the study team and other departments of the institution.

9.3 Well-Designed CRF

Informative header and footer is an important component of well-referenced CRF. Typical headers and footers contain information such as Sponsor ID, Protocol number, Subject ID, Subject initials and page number, all of which uniquely identify a CRF page. Although these fields are sufficient in managing data, additional information such as protocol label, visit ID and label, CRF version and date printed, as well as sequence number make database design easier and a better link to the study database.

The CRF version number is a critical field that not only prevents using an incorrect CRF page but also confirms that there is no change made to the page. If a different version number is used, it warns users about possible changes within the page and the database should be redesigned to accommodate the change.

It is essential to organize CRF pages in a structural order, reflecting the schedule of assessments specified in the protocol. A table of contents acts as a reference for the sequential order of CRFs. Not only does it provide site personnel with a quick reference to specific pages (for recording the study data) but it also helps define the database in a structural manner.

9.4 CRF Design Layout

There are three types of data: non-time dependent, time dependent, and cumulative data.

Non-time dependent data: Non-time dependent data is the data collected at a snapshot in time. Such data include subject demographics and medical history.

Time dependent data: Time dependent data is data collected repeatedly over time. A typical example is vital signs recorded at multiple visits. With time dependent data, there are 2 options to the CRF layout: Single page, per visit or a cumulative log. With the first approach, the data is represented at each visit while the second approach is a single page with multiple records representing the "repeated" time measurements. Both approaches have pros and cons. The "pervisit" approach more accurately reflects the schedule of assessments but could lead to a larger

CRF booklet: The "cumulative log" approach saves number of pages in a CRF booklet and makes variables structured in the same way as the CRF page in the database. However, since it does not allow the CRF retrieval as "per-visit", it may be inconvenient for investigators to frequently flip over pages from the log section to the actual visit section. It also restricts retrieval of data for a plan of data review by scheduled visits. Furthermore, it is more likely to yield data entry errors if too many fields are combined into one page and become cluttered. The "per-visit" approach is preferred for assessments such as physical examination or laboratory data as they involve many parameters. The "cumulative log" approach may be preferred for groups of assessments such as vital signs, which involve a fewer number of parameters. Cumulative data: Cumulative data is data collected over time but not linked to a specific visit. Adverse events and concomitant medications are typical examples. The usual approach to designing a CRF for cumulative data is the "cumulative log" approach described in the previous section.

9.5 CRF Field Designs

Well-designed CRFs use different types of icons for different response formats. This helps obtain consistency in data format. They provide a clear picture of what is expected in the field for investigators and coordinators to help report valid data.

A pre-coded response (i.e. 1 = Mild, 2 = Moderate, or 3 = Severe) is primarily used to aid data entry and statistical analysis. It makes data manipulation easier during a statistical analysis process.

Providing an example and expected format for a field can reduce data misinterpretations and also reduce the number of un-necessary queries that would be issued to clarify the data. For instance, a date, written as "02/03/05", for instance, can be misread as either "March 02, 2005" or "March 05, 2002" when there is no specification of date format. This affects the validity of data. In order to avoid this, the date format, "dd/MMM/yyyy", can be used as shown in the Figure 5. Entering data with correct format assists in establishing validity and consistency of the data. Unnecessary data queries, interpretation and data manipulation can be reduced during the data validation and statistical analysis processes.

9.6 CRF Verification Before Use

Case report form should be verified for the following:

I. Order of the information recorded

II. Missing modules and pages

III. Changes in the page numbering during the successive versions

IV. The size of space for answers
V. The content (a 100% correspondence between CRF and protocol)
VI. Absence of obscure and ambiguous questions

9.7 Elements of Trial Design

CRF completion guidelines can be inserted to provide study-specific data collection procedures under standard regulatory guidelines. Special instruction pages for filling out the CRF should be planned. Data formats for appropriate response fields, a data correction guide, how to handle unknown or unavailable data, and a retrieval schedule for completed CRFs can be outlined. For instance, a procedure for completing subject initials in the header can be included. The guidelines can remind investigators that their initials cannot be changed even if a subject's name is changed as a result of marriage or divorce during the study. They can also instruct how to handle missing or unavailable data. If a required piece of information or entire section cannot be retrieved, the use of "NA", "ND" or "UNK" can be defined to avoid ambiguous responses. To handle an unknown or unavailable date response, an imprecise date format can be suggested in the guideline. When 'day' and/or 'month' of date are unknown, unknown day and month format can be defined as "UK/UNK/2000" and forms that do not allow an imprecise date (eg. Study Drug Dosing) can be listed. Moreover, data correction rules can be specified such as drawing a single line through the original entry with a signed, dated correction. The guidelines can also demonstrate how to complete CRF pages for unscheduled assessments or in the cumulative log. Furthermore, CRF retrieval procedures including how to handle CRFs for subjects who have discontinued during the study can be listed. In order to enhance the efficiency of CRF completion procedures, there should be a study-specific CRF completion guidelines.

Completion-validation of the Case Report Form

The Investigator must ensure the accuracy, legibility and completeness of data entry in the case report forms (CRFs) and in all other required report forms/logs. All CRFs and other required forms will be validated by the Clinical Monitor during Monitoring Visits.

Completion of CRF

1. Only authorized study staff (name shown in authorized signatory form - ASF) are allowed to enter data into the CRF and other required report forms.
2. Ball-point pen must be used.
3. Capital letters must be used for all entries in the CRF.
4. All items must be completed by entering a number or text in the space provided.

5. When a subject is recruited into the trial, the initial and allocated numbers are entered in the CRF against the subject's name on the Subject Identification Code List. The subject's name should never be entered in the CRF to preserve confidentiality.
6. As far as possible, the results of assessment should be first be entered into the subject file and then transcribed into the CRF. This will allow data to be verified during the process of source data verification.
7. The CRF should be completed during subject participation.
8. Data reported on the CRFs that are derived from source documents should be consistent with the source documents or the discrepancies should be explained.

Case Report Form Corrections

1. Only authorized study staff can make corrections.
2. Clinical Monitor/Sponsor should not be allowed to make correction in the CRF.
3. Corrections should not obscure the original entry:
   - do not erase
   - do not overwrite
   - never use correcting fluids.
4. To make a correction:
   - cross out the wrong entry with a single line.
   - write the correct entry alongside/above/under the wrong entry.
   - date the correction
   - initial the correction
   - explain the correction (if necessary).

Example:

~~PAZ~~ CL

PAT 9/11/98

Sex ~~1~~

Male: 1 2 PAT 9/11/98 ?

Female: 2

Date of vaccination ~~2/10/95~~

2/10/98 PAT 9/11/98

9.8 Electronic Data Entry Forms

When clinical trials rely heavily on paper documents, data collection and management consumes substantial resources in time and effort. There are many reports which suggest that the efficiency of clinical trial can be improved by replacing paper CRF with eCRF. With internet based applications in many fields, the demand of eCRF is growing.

An eCRF or electronic CRF is a set of computer forms which are based on the traditional paper case report forms. Using a standard wave browser these forms are accessible over a secured internet connection.

An Electronic Data Capture (EDC) system is a computerized system designed for the collection of clinical data in electronic format for use mainly in human clinical trials.

Typically, EDC systems provide:

- a graphical user interface component for data entry
- a validation component to check user data
- a reporting tool for analysis of the collected data

An EDC comes as a remote data entry device which does not need public or private network infrastructure. Most common EDC appearance is on laptop with preconfigurated software for a single trial which is distributed to study site.

Developing Regulatory compliant eCRFs – The principles of GCP for data integrity must apply equally to EDC in similar lines of paper based data collection. The use of computerized systems developed for clinical trials are specially regulated. The electronic records, Electronic signature regulation, known as 21CRF part 11, addresses issues for quality, security and integrity of data which is accepted as equivalent to paper records. FDA republished in 2007, a guidance for computerized systems used in clinical trials, while in Europe the requirement for eCRF are included in the GCP guideline of the EMEA published in 2002.

To achieve regulatory compliance, system and research projects employ procedural and technical controls. Procedural controls involve administrative actions, dealing with trial organization and documentation, while technical controls comprise the measures to ensure the quality, accuracy and integrity of data stored in electronic system. To prevent unauthorized access to system, user authorization controls are used to identify the person who submits the data as a security measure. Audit trail controls are measures to ensure that system keeps a record about sources from which data originates, who made changes, when, and what information was changed. Attributability controls are a measure to ensure that the data will be retrievable in such a way that all information regarding each subject is attributable to that subject. Data validation controls are checks performed by computerized systems to ensure the validity and quality of clinical information. System integrity measures are all the procedures to guarantee the integrity of the system and protect against data loss.

9.9 Conclusion

A well-designed CRF booklet can help define a structured database and collect valid and consistent data in a clinical trial. It reduces time on data query management and increases the efficiency of statistical analysis and output generation. The layout of a CRF should be determined by data types and the structure of the dataset in the database while considering a cluster level of fields in one page as well as a plan of data review by scheduled visits. The well-organized CRF fields give users a clear direction of required responses. The response format should be specified in the field to ensure accurate data transcription and consistency. Finally, the CRF completion guidelines which contain detailed instructions for completing and correcting CRFs, signing procedures, and handling of completed CRFs under the industry standards must be followed. All of these techniques and considerations contribute in developing well-designed CRFs under "best practices".

9.10 Points to Remember

The data collected in the study site during the course of a study should be comprehensive and provide true and fair information on what happened to each study subject. Only if the above criteria are met, the study will reliably answer the questions concerning the efficacy and safety of the investigational drug.

All CRF's should include the following data:

- study title and number;
- Investigator's name;
- study subject/patient ID (number and initials);
- inclusion / exclusion criteria;
- demographic data;
- detailed description of dosage regimens of investigational drug;
- concomitant treatment;
- adverse events (side effects and intercurrent diseases);
- conclusion on subject's health;
- Investigator's signature and date.

Additionally, the CRF's should include special pages to record the following information:

- past medical history;

- results of physical examination;
- primary and secondary diagnoses;
- relevant previous treatment;
- baseline characteristics, results of interim assessments, evaluation of efficacy endpoints, laboratory tests, description of study procedures etc.

Best Practices

- Design the CRF along with protocol to assure collection of only the se data the protocol specifies.
- Keep questions, prompts and instructions clear and concise.
- Design the CRF to follow the data flow from the perspective of the person completing it, taking into account the flow of study procedures and typical organization of data in a medical record.
- Avoid referential and redundant data points within the CRF whenever possible. If redundant data collection is used to assess data validity, the measurements should be obtained through independent means.
- Design the CRF with the primary safety and efficacy endpoints in mind as the main goal of data collection.
- Establish and maintain a library of standard forms.
- Make the CRF available for review at the clinical site prior to approval.
- Use NCR (no carbon required) paper or other means to assure exact replicas of paper collection tools.

Suggested Reading

1. ICH Guidance E6: Good Clinical Practice: Consolidated guideline
2. Cochrane Collaboration Glossary
3. M.V. Leonova, I.L. Asetskaya, Development of Study Protocol and Case Record FormRussian State Medical University, Moscow. Article in Russian language.
4. T.K. Efimtseva, A.L. Spasokukotsky, O.V. Mironova. Good Clinical Practice. English-Russian glossary. Ukrainian medical Journal - No 2 (28) - III / IV 2002.
5. Shein Chung Chow. Design and analysis of clinical trials. (Concepts and methodologies). Wiley Publication
6. Bogdan Ene-Iordache et al. Developing regulatory compliant Electronic case report forms for clinical trials: experience with the Demand trial. J Am Med Inform Assoc 2009; 16: 404-408.

CHAPTER 10

Process of Randomization in Clinical Trial

M.U.R. Naidu and P. Usha Rani

10.1 Introduction

In clinical trial, randomization is a process by which each trial participant has the equal chance of being assigned to either active intervention or control group. Randomization is the basis for statistical inference, and is probably the best method for achieving comparability. With respect to both known and unknown risk factors, in any clinical trial, randomization tends to produce comparable study groups, removes the bias in allocating the trial subjects and provides a guarantee that statistical tests will have true, valid significance levels. In the process of randomization, neither the investigator nor the subject influences the treatment assignment. Random selection or assignment does not guarantee the representativeness of the sample, but it does reduce the bias.

To eliminate experimental bias, several methods of randomly allocating trial subjects in clinical trials are used. There are mainly two types of experimental biases which are important, one is "Selection bias" which occurs due to predictable allocation process. In this, the decision to assign a participant to a trial may be influenced by the anticipated treatment assignment. Here a selection bias will occur if any bias exists as to what treatment a particular type of trial subjects should receive. Most of the processes of randomization avoid selection bias by not being predictable.

The second type of bias is called "Accidental bias" which can occur if the clinical trial does not achieve balance on risk factors or prognostic covariates. Usually some randomization procedures are more vulnerable to accidental bias especially for small sample size clinical trials. Thus, it is important, that every trial report must clearly describe the randomization procedure used. Description of randomization should be brief, clearly indicating type of randomization method and how it is implemented.

10.2 Advantages of Randomization

Using randomization in a controlled clinical trial has the following advantages:

1. It attempts to decrease investigator bias in assigning the subject to treatment groups.
2. Randomization permits certain statistical methods to be used with the resulting data.
3. It allows for blinding of subjects and investigator in such a manner that if the blind is broken for one patient, it is not broken for all.
4. For demonstrating efficacy and safety of investigational medicine, the randomization procedures are the current norms.

Limitation of using Randomization in clinical trials:

1. Randomization alone cannot adequately eliminate the differences in the treatment groups if the baseline values were widely divergent.
2. Randomization may not assure an equal distribution of risk factors or other relevant patient characteristics to each treatment group.
3. Administration of randomization process into a clinical trial may increase the complex or difficulty which may pose practical problems in analyzing the data.

10.3 Types of Randomization

(a) **Fixed Allocation Randomization Procedure:** In this procedure, the intervention is assigned to the trial subjects with pre-specified probability like equal and as the study progresses the probability of allocation is unaltered. Several methods can be used for fixed allocation like

(i) Simple randomization

(ii) Block randomization

(iii) Stratified randomization

It is desirable that the allocation to intervention and control groups should be equal. In some clinical trials the unequal allocations such as 2:1 or 3:1 interventions are used. In unequal allocation, though the study may gain more information of intervention response like efficacy or side effects, it may loose the sensitivity. In this allocation less information is needed about control group and therefore fewer control subjects are required. In unequal allocations, sometimes there is the risk of new interventions turning harmful, in which case more trial subjects are exposed to a toxic drug.

(i) ***Simple Randomization:*** Simple randomization is also known as complete randomization, and is the most elementary form of randomization. A random digit table on which equally likely digits 0-9 are arranged in rows, and columns are usually used to generate simple randomization, this method is useful for small studies. By randomly selecting a certain row and observing the digits in that row, one can assign even digit to group "A" and odd digit to Group "B". This method provides a sequence of assignments that is random in order and each trial subject has an equal chance of being assigned to "A" or "B" group.

A random number producing algorithm available on computer can be used to generate randomization schedule for large studies. Computerized process for simple randomization is to use a uniform random number algorithm to produce random number from 0.0 - 0.999. This procedure can be applied easily to more than two groups. For example, the trial has four groups A,B,C and D and participants are to be randomized such that a participant has a one-fourth chance of being in group "A", "B", "C" or "D". By dividing the interval 0 to 1 into four pieces of length 1/4th, 1/4th, 1/4th and 1/4th, random number generated will have probability of 1/4th in each group. The interval would be 0 to 0.249, 0.25 to 0.499, 0.50 to 0.749 and 0.75 to 0.999. Random number falling between 0-0.249 is assigned "A" and 0.25 – 0.499 "B", 0.50 – 0.749 "C' and from 0.75 – 0.999 "D". The advantage of simple randomization procedure is that it is easy to implement. The disadvantage is that there can be imbalance. Primarily for this reason simple randomization is not usually preferred even for large studies.

(ii) ***Block Randomization:*** Another desirable restriction in the treatment assignment process may involve blocks, which results in an equal number of subjects in each treatment with multiple blocks based on the number of treatments. In process of forming blocks, many subtle factors that might influence the important outcome measurements may change over time. This was first described by

Hill in 1951. This method is used to prevent the occurrence of imbalance in participant's assignment to various groups. At no time during the randomization, the allocation will be imbalanced if the block randomization is properly applied. Trial subjects are assigned randomly with equal probability to each group A or B, one half of the subjects is assigned "A" and other half to "B". The order of intervention in each block is randomized. Each block is of even size (eg. 4 ,6, or 8). In a block size of 4 subjects there can be 6 possible combinations of assignments. ABAB, AABB, BAAB, BABA, BBAA and ABBA. The advantage of blocking is that if the trial is terminated before the enrollment is completed, balance will exist in terms of the number of subjects randomized to each groups.

A potential problem with this type of randomization is that, if the blocking factor is known, and study is not double blind, then the assignments of last person in each block is known eg if the block factor is 4 and the first 3 assignments are ABB, then the next must be A. This could introduce a bias, thus the blocking factor must be unknown to the study team. One of the disadvantages of blocked randomization is that, the analysis of data is more complicated than simple randomization because data analysis requires the information on randomization process actually performed in block randomization for statistical analysis. Since blocking guarantees balance between the two groups, this increases the power of the study.

(iii) ***Stratified Randomization:*** The investigator may wish to use stratification, a process that divides subjects into subsets or strata, to make sure that each stratum is represented in the sample so that the treatment groups are comparable at the beginning of the study. At baseline certain characteristics like prognostic or risk factors need comparability between the groups. Various factors that correlate with subsequent response or outcome, if not evenly distributed between the intervention and control group will affect the trial results. For any simple study, there is no guarantee that all baseline characteristics will be similar in the two groups. Stratified randomization is a method that helps to achieve comparability between the study groups. This type of randomization requires that the factors be measured at the time of randomization. If several factors are used, a stratum is formed and one subgroup from each of them is selected. Randomization of participants is done within the stratum, which can be a simple randomization.

Stratified randomization breaks the trial down into smaller subgroups and participants in each of the smaller groups belong to the same stratum. This reduces variability in group comparisons if the stratification is used in the analysis. Reduction in variability allows the detection of specified differences with fewer participants. For a large study group, stratification either before or after randomization provides equal efficiency. However, for studies of 100 participants or fewer, the stratification before randomization may achieve greater power. For all potential problems of baseline imbalances, stratified randomization is not a complete solution.

In multicenter trial, the subject recruitment type, quality of care may vary from center to center, hence it is advisable to do the randomization process stratified accordingly. Another reason for centre wise stratification is that, if a center is eliminated from the study, the balance in other centers will not be affected.

(b) **Adaptive Randomization Procedure:** In contrast to fixed allocation, adaptive procedure changes the allocation probabilities as the study progresses. There can be two types of adaptive allocations, one adjusts the allocation depending on imbalance in number of subject or baseline characteristics between the two groups, and the other adjusts allocation according to the response to the assigned intervention.

(i) ***Baseline Adaptive Randomization:*** The purpose of the algorithm is basically to randomize the allocation to group A and B with equal probability, as long as the number of subjects in each group is equal or near equal. In case, if imbalance occurs and the difference in number of subjects is greater than the pre specified value, the allocation probability is adjusted, so that the probability is higher for the group with fewer subjects. This procedure can be modified with consideration of the number of consecutive assignments to the same group and the length of such a run. The biased coin randomization procedure, attempts to balance the number of participants in each treatment group based on the previous assignments, but does not take participants responses into consideration. This approach demands a cumbersome data analysis. One possible advantage of the biased coin approach over the block randomization is, that the investigator cannot determine the next assignment by discovering the blocking factor. Because of the complexity, the biased coin method is not widely used as the blocked randomization.

Another adaptive randomization method is referred as the Urn design. As the trial progresses, this method also attempts to keep the number of subjects randomized to each group reasonably balanced. Like the biased coin design, it also requires the analysis to account for the randomization.

Other adaptive stratification methods consider the intervention assignment probabilities as a function of the distribution of the prognostic factors for subjects who are already randomized. For example, if age is a prognostic factor and one group "A" has more older subjects, than the other group "B", the allocation scheme is such that, the next several older subjects would most likely be randomized to group "B". Generally in adaptive stratification method as several prognostic factors are incorporated overall assessment of group balance is done and subjects are then assigned to that group to correct the imbalance. This method is also called "minimization".

Minimization gives unbiased estimates of treatment effect thereby increasing the power of the study than stratified randomization. One of the major advantages of this procedure is that it protects against a severe baseline imbalance for important prognostic factors. Adaptive stratification is operationally more difficult to carry out.

(ii) **Response Adaptive Randomization:** To allocate the next participant, response adaptive randomization uses information on participant response to intervention. This model assumes that the investigator randomizes subjects to one group, of two interventions and the primary response variable is determined quickly relative to total duration of study. The "play and winner procedure" may assign the first patient by the toss of a coin; the next patient is assigned to the same group as the first patient if the intervention response is successful, otherwise the patient is assigned to other group. The other response adaptive procedure is the two-armed "bandit" method, which continually updates the probability of success as soon as the outcome for each participant is known. The information is used to adjust the probability of assignment to either group, so that high proportion of future participants would receive the currently better or more successful intervention. Both methods above have the intended purpose of maximizing the number of participants to superior intervention.

Although these methods do maximize the number of subjects to superior intervention, imbalance will result in loss of power and require more participants to be enrolled into the study. Analysis is usually more complicated with response adaptive methods than with simple randomization. Due to these disadvantages, response adaptive procedures are not used commonly.

10.4 Techniques of Randomisation

Although historically "manual" randomization techniques (such as shuffling cards, drawing pieces of paper from a bag, spinning a roulette wheel) were common, nowadays automated techniques are mostly used. As both selecting random samples and random permutations can be reduced to simply selecting random numbers, random number generation methods are now most commonly used, both hardware random number generators and pseudo-random number generators.

Physical methods include:

- Casting yarrow stalks (for the I Ching)
- Throwing dice
- Drawing straws
- Shuffling cards
- Roulette wheels
- Drawing pieces of paper or balls from a bag
- "Lottery machines"
- Random Tables:

For example: Randomization Table

- Select an arbitrary starting point—

  i.e., first line, 5th number:

- 54 99 20 22 52 49 06 78 58 51 62 85 55

  12 88 33 60 26 13 69 74 80 71 48 70 73

  19 62 58 20 50 59 08 67 01 ————

  Give Treatment A for odd number and Treatment B for even number

Randomization Scheme

Subject	Random No	Group
1	52	B
2	49	A
3	06	B
4	78	B
5	58	B
6	51	A
7	62	B
8	85	A
9	55	A

Random Allocation Software

Random allocation software has been produce to support the first type of randomization, i.e. block randomization. This software produces as its output a sequence of allocation based on the selected type of blocking. Below is the description of this software.

Random Allocation Software is a Windows software which install in the same way as ordinary windows software (i.e. running setup.exe and following on screen instructions). After installing and running the software the main window will appear which is as follow.

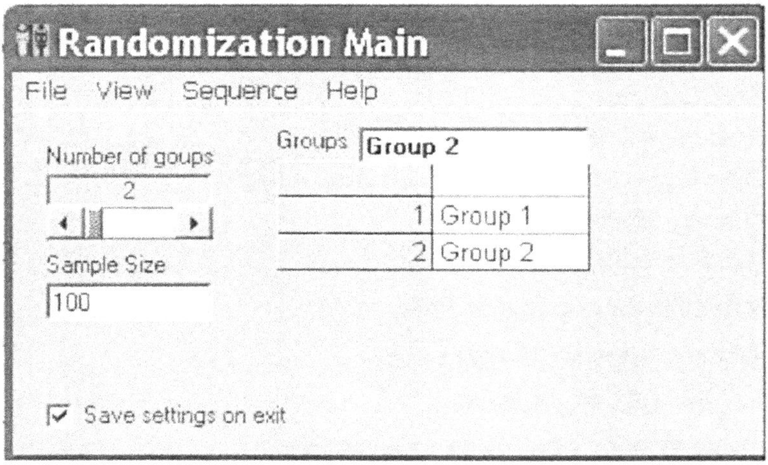

Steps to generate a random sequence

1. Select the number of groups (2 - 16)
2. Enter the name of each group
3. Enter the sample size
4. Select 'Generate' from the 'Sequence menu'

(Random Allocation Software has been developed by M. Saghaei, MD., Department of Anesthesia, Isfahan University of Medical Sciences, Isfahan)

A **randomized algorithm** or **probabilistic algorithm** is an algorithm which employs a degree of randomness as part of its logic. The algorithm typically uses uniformly random bits as an auxiliary input to guide its behavior, in the hope of achieving good performance in the "average case" over all possible choices of random bits. Formally, the algorithm's performance will be a random variable determined by the random bits; thus either the running time, or the output (or both) are random variables. In common practice, randomized algorithms are approximated using a pseudorandom number generator in place of a true source of random bits; such an implementation may deviate from the expected theoretical behavior.

Central Randomisation: IVRS / IWRS

Multi-site, multi-country clinical trials can be difficult to manage. However, an **Interactive Voice Response System (IVRS)** combined with ubiquitous phone and fax services, or an **Interactive Web Response System (IWRS)** using a standard web browser and email service, allows study administrators and investigators to securely interact with the study database, making study deployment fast and easy. Normally both these services can be accessed from any location worldwide.

When used for Patient Randomisation, the IVRS allows study administrators and investigators to interact with a computer database using a telephone touchtone keypad as a means of data input. The computer is able to prompt the caller for required information with straightforward pre-recorded voice messages (in any language) and give out information in response.

Similarly, the IWRS assigns treatments to subjects, provides automated confirmation of assignments (on screen and via email), as well as regular reporting of trial progress to the researcher. With online interactions the IWRS system expands on the depth of detail that an investigator or administrator is able to obtain about the progress of the trial compared to that available through IVRS.

Both services randomise treatments to subjects, provide automated confirmation of assignments, as well as regular reporting of trial progress to the researcher. All this is done on a 24/7 basis, eliminating the issues of time-zones and distance.

Benefits

IVRS and IWRS have gained wide acceptance because they provide real-time benefits in Patient Randomisation, drug assignment and inventory management, study blinding, reporting, systems integration and low costs:

- Randomisation. Real-time feedback from a centralised database allows a researcher to dynamically allocate treatments to subject groups to better aid statistical analysis.
- Drug Assignment. Knowing where and when to allocate expensive trial drugs saves valuable resources.
- Study Blinding. IVRS/IWRS helps improve the blinding process by allocating treatment groups in a controlled and unbiased manner.
- Reporting. Through automated system reports, real-time information can be made available to the study manager via fax and email.
- Systems Integration. The IVRS/IWRS and the initial study preparations and documentation fit into the familiar manual processes of organisations running clinical trials.

- Low Cost. Provided as a hosted service there is no capital expenditure required - you only pay an initial study setup fee and a low monthly fee for the provision of Services for the duration of the study.

Disadvantages of using IVR System

The greatest disadvantage of IVR systems is that many people simply dislike talking to machines. Older adults may have a hard time following telephone menus and lengthy instructions. And younger callers get frustrated with the slowness of multiple phone menus.

The problem with IVR systems is that it's hard to design a good hastle free system. Some of the most common user complaints with IVR systems: include the following:

- Menus are too long. Experts recommend that no menu should exceed four choices. This makes it easy to remember the options and doesn't waste the caller's time listening to tons of choices.
- There's too much information. Hence it is better to develop a simple system and wait for a caller to ask for help or request more information instead of offering it all up front.
- Voice prompts are hard to understand. This could be caused by two different factors. To save money, the organization didn't hire professional voice talent and may have recorded the audio over the phone instead of in a studio. Or, if the organization opted to use an automated voice, they may have chosen cheap text-to-speech software that's hard to understand.
- The accent may be difficult to understand especially if the caller is from a non-english speaking state.

10.5 Implementation of Randomization Procedure

It is extremely important to pay careful attention for implementing the actual randomization process. Improper selection of randomization will lead to compromising of the advantages of using the process.

Independent central unit should develop randomization process and assign participants to the appropriate groups. This is highly recommended for valid randomization process in any study. Randomization schedules are made before the study begins, in many cases when a fixed proportion randomization process is used, investigator calls the center for assignment of next participant. Another possibility is to have a sealed envelop containing the assignments. As the participant enters the trial, he or she receives the next envelope. In an other procedure used in double blind trial, medication bottles are numbered, and distributed to participants in sequence.

Many studies using a fixed proportion randomization schedule require investigator to call randomization center to verify that a subject is eligible to be enrolled before any assignment, this allows only that eligible participants will be randomized.

The randomization should be in blocks for studies involving more than several hundred participants. In a large multicentric trial, randomization and stratification should be centralised. For studies with small sample size, randomization should be in blocks and stratified to assure that balance is achieved with respect to important risk factors.

Suggested Reading

1. Spilker B. Guide to clinical trials. New York 1991. Raven Press
2. Friedman LM, Furberg CD and DeMets DL. Fundamentals of clinical trials, 3rd ed. Springer; New York 1998.

Chapter 11

Investigational Medicinal Product

Y.Sridhar, Abdul Mateen, P.Usha Rani and M.U.R.Naidu

11.1 Definition

Authorized Prescriber

An authorized provider is one who is authorized to prescribe the IMP as a Principal Investigator (PI) or Site Investigator (SI).

Blind

The term blind refers to the lack of knowledge regarding the identity of the drug treatment. In a clinical trial, patients, investigators, and other personnel may be kept "blinded" to the medication prescribed in order to decrease biases and avoid placebo effect.

Reference product

A reference product is an agent that the IMP is being compared to in a clinical trial. A reference drug may be the current standard of care for the disease state being studied.

Centralized Dispensing Protocol

A centralized dispensing protocol is a protocol that requires dispensing of the IMP and possibly concurrent or reference medications from a single centralized\ pharmacy, directly to the study participants.

ooperative Study

. cooperative study is a project or program of research at two or more health are facilities which utilizes a common protocol, so that data obtained at all articipating facilities can be treated as though from a single source.

nvestigational Supply

n investigational supply is any medical or surgical product listed in the National rug File and is not considered a drug. These products are identified as epartment of Veterans Affairs (VA) Class IN-000, and are not subject to co-ayment.

epass or Retest Date

. repass or retest date is the date assigned by the manufacturer after which the rug substances need to be examined to ensure that they are within suitable pecifications for use in the manufacture of a drug product.

nvestigational Medicinal Product (IMP)

n IMP is a chemical or biological drug that is used in a clinical investigation.

(a) A new chemical compound, which has not been released by the Food and Drug Administration (FDA) for general use, or

(b) An approved drug that is being studied for an approved or unapproved use, dose, dosage form, administration schedule, or under an Investigational New Drug (IND) application, in a controlled, randomized, or blinded clinical trial.

Concurrent medications, comparators, or rescue medications used in the ivestigational trial that are not the drug(s) being studied are not defined as IMP nless they are not commercially approved or not available through commercial hannels.

Prescription drugs, over-the-counter drugs, nutritional supplements, herbal reparations, and legend items used for diagnosis or treatment are considered s IMP.

lanning trials – Precautions to be Taken

he sponsor should ensure that sufficient safety and efficacy data from preclinical udies and/or clinical trials are available to support human exposure by the oute, at the dosages, for the duration, and in the trial population to be studied. he sponsor should update the Investigator's Brochure as significant new iformation becomes available.

11.2 Manufacturing, Packaging, Labeling, and Coding IMPs

The sponsor should ensure that the IMP's (including active comparator(s) and placebo) is characterized as appropriate to the stage of development of the product(s), is manufactured in accordance with any applicable GMP, and is coded and labelled in a manner that protects the blinding (if applicable). In addition, the labeling should comply with applicable regulatory requirement(s).

The sponsor should determine, for the IMP's, acceptable storage temperatures, storage conditions (e.g., protection from light), storage times, reconstitution fluids and procedures, and devices for product infusion, if any. The sponsor should inform all involved parties (e.g., monitors, investigators, pharmacists, storage managers) of these determinations.

The IMP's should be packaged to prevent contamination and unacceptable deterioration during transport and storage.

In blinded trials, the coding system for the IMP's should include a mechanism that permits rapid identification of the product(s) in case of a medical emergency but does not permit undetectable breaks of the blinding.

If significant formulation changes are made in the investigational or comparator product(s) during the course of clinical development, the results of any additional studies of the formulated product(s) (e.g., stability, dissolution rate, bioavailability) needed to assess whether these changes would significantly alter the pharmacokinetic profile of the product should be available prior to the use of the new formulation in clinical trials.

11.3 Supply and Handling IMPs

The sponsor is responsible for supplying the investigator(s)/institution(s) with the IMP's. The sponsor should not supply an investigator/institution with the IMP's until the sponsor obtains all required documentation (e.g., approval, favourable opinion from IRB/IEC and regulatory authorities). The sponsor should ensure that written procedures include instructions that the investigator/institution should follow for the handling and storage of IMP's for the trial and documentation thereof. The procedures should address adequate and safe receipt, handling storage, dispensing, retrieval of unused product from subjects, and return of unused IMP's to the sponsor (or alternative disposition if authorized by the sponsor and in compliance with the applicable regulatory requirements).

The sponsor should

(a) Ensure timely delivery of IMP's to the investigator(s).

(b) Maintain records that document shipment, receipt, disposition, return and destruction of the IMP's.

(c) Maintain a system for retrieving IMP's and documenting this retrieval (e.g., for deficient product recall, reclaim after trial completion, expired product reclaim).

(d) Maintain a system for the position of unused IMP's and for the documentation of this disposition.

(e) Take steps to ensure that the IMP's are stable over the period of use.

(f) Maintain sufficient quantities of the IMP's used in the trials to reconfirm specifications, should this become necessary, and maintain records of batch sample analyses and characteristics. To the extent stability permits, samples should be retained either until the analyses of the trial data are complete or as required by the applicable regulatory requirement(s), whichever represents the longer retention period.

afety Information

'he sponsor is responsible for the ongoing safety evaluation of the IMP's. The ponsor should promptly notify all concerned investigator(s)/institution(s) and ie regulatory authority of findings that could affect adversely the safety of ubjects, impact the conduct of the trial, or alter the IRB/IEC's approval/ ıvourable opinion to continue the trial.

1.4 Receipt of IMP, Inventory, Storage and Accountability

eceipt of IMP

'he Investigator / designee site pharmacist has the following responsibilities egarding IMP's:

- Maintain records of IMP's receipt to the study site, regarding delivery dates, quantities received, batch/serial numbers, and expiration dates.
- Maintain an inventory of the IMP's at the study site.
- Inventory control records should be updated, signed, and dated by the investigator in a timely manner.
- Record/track use of the IMP's by each subject.
- Documentation should verify that dosing use was in accordance with the approved protocol. Maintain an accountability log that records, when the subject(s) received the IMP and the specific dosage the subject(s) received.
- Return/dispose of unused IMP's as specified by the sponsor, maintaining documentation of return/disposal.

torage the IMP's

he storage area should be locked/secure with access limited to approved study harmacist only. Drugs/devices should not be stored with standard clinical ıventory.

Accountability of IMP's

Responsibility for IMP accountability at the trial site(s) rests with the investigator or institution. The investigator/institution should assign for IMP's(s) accountability at the trial site(s) to an appropriate pharmacist or another appropriate individual who is under the supervision of the investigator/institution.

The investigator or person designated by the investigator/institution, should maintain records of the IMP's delivery to the trial site, the inventory at the site, the use by each subject, and the return to the sponsor or alternative disposition of unused IMP's. These records should include dates, quantities, batch/serial numbers, expiration dates and the unique code numbers assigned to the IMP's and trial subjects. Investigators should maintain records that document adequately that the subjects were provided the doses specified by the protocol and reconcile all IMP's received from the sponsor. The IMP's should be stored as specified by the sponsor and in accordance with applicable regulatory requirement's. The investigator should ensure that the IMP's are used only in accordance with the approved protocol. The investigator, or a person designated by the investigator/institution, should explain the correct use of the IMP's to each subject and should check, at intervals appropriate for the trial, that each subject is following the instructions properly.

11.5 Randomization Code

Procedures should describe the generation, security, distribution, handling and retention of any randomisation code used for packaging IMP's, and code-break mechanisms. Appropriate records should be maintained.

The investigator should follow the trial's randomization procedures, if any, and should ensure that the code is broken only in accordance with the protocol. If the trial is blinded, the investigator should promptly document and explain to the sponsor any premature unblinding (e.g., accidental unblinding, unblinding due to a serious adverse event) of the IMP's(s).

11.6 Principles Applicable to Comparator Product

If a product is modified, data should be available (e.g. stability, comparative dissolution, bioavailability) to demonstrate that these changes do not significantly alter the original quality characteristics of the product.

The expiry date stated for the comparator product in its original packaging might not be applicable to the product where it has been repackaged in a different container that may not offer equivalent protection, or be compatible with the product. A suitable use by date, taking into account the nature of the product, the characteristics of the container and the storage conditions to which the article may be subjected, should be determined by or on behalf of the sponsor. Such a date should be justified and must not be later than the expiry date of the original package. There should be compatibility of expiry dating and clinical trial duration.

Packaging

During packaging of IMP's, it may be necessary to handle different products on the same packaging line at the same time. The risk of product mix up must be minimised by using appropriate procedures and/or, specialized equipment as appropriate and relevant staff training.

Packaging and labeling of IMP's are likely to be more complex and more liable to errors (which are also harder to detect) than for marketed products, particularly when "blinded" products with similar appearance are used. Precautions against mislabeling such as label reconciliation, line clearance, in process control checks by appropriately trained staff should accordingly be intensified.

The packaging must ensure that the IMP's remains in good condition during transport and storage at intermediate destinations. Any opening or tampering of the outer packaging during transport should be readily discernible.

Investigational Product Labeling

Document whether there is precise labeling that ensures that the product reaches the proper clinical site if more than one site is involved in the study. In addition, there should be documentation included in the review that describes product labeling throughout the manufacturing process. You should verify that any proposed labeling contains the date of product manufacture, storage conditions, expiration date and possibly time, product name, and patient identifiers. Two unique patient identifiers should be used to minimize the potential for any mix-ups. In addition, as described in 21 CFR 312.6, the label for an IMP's must contain the following statement: "Clinical Research Purpose Only".

The following information should be included on labels, unless its absence can be justified, e.g. use of a centralized electronic randomization system:

(i) Name, address and telephone number of the sponsor, contract research organization or investigator (the main contact for information on the product, clinical trial and emergency unblinding);

(ii) Pharmaceutical dosage form, route of administration, quantity of dosage units, and in the case of open trials, the name/identifier and strength/potency;

(iii) The batch and/or code number to identify the contents and packaging operation;

(iv) A trial reference code allowing identification of the trial, site, investigator and sponsor if not given elsewhere;

(v) The trial subject identification number/treatment number and where relevant, the visit number;

(vi) The name of the investigator (if not included in (i) or (iv));

(vii) Directions for use (reference may be made to a leaflet or other explanatory document intended for the trial subject or person administering the product);

(viii) "For clinical trial use only" or similar wording; The storage conditions;

(ix) Period of use (use-by date, expiry date or re-test date as applicable), in month/year format and in a manner that avoids any ambiguity.

(x) "Keep out of reach of children" except when the product is for use in trials where the product is not taken home by subjects.

The address and telephone number of the main contact for information on the product, clinical trial and for emergency unblinding need not appear on the label where the subject has been given a leaflet or card which provides these details and has been instructed to keep this in their possession at all times.

Symbols or pictograms may be included to clarify certain information mentioned above. Additional information, warnings and/or handling instructions may be displayed. If it becomes necessary to change the use-by date, an additional label should be affixed to the IMP. This additional label should state the new use-by date and repeat the batch number. It may be superimposed on the old use-by date, but for quality control reasons, not on the original batch number. This operation should be performed at an appropriately authorised manufacturing site. However, when justified, it may be performed at the investigational site by or under the supervision of the clinical trial site pharmacist, or other health care professional in accordance with national regulations. Where this is not possible, it may be performed by the clinical trial monitor(s) who should be appropriately trained. The operation should be performed in accordance with GMP principles, specific and standard operating procedures and under contract, if applicable, and should be checked by a second person. This additional labelling should be properly documented in both the trial documentation and in the batch records.

11.7 Blinding Operations

Where products are blinded, systems should be in place to ensure that the blind is achieved and maintained while allowing for identification of "blinded" products when necessary, including the batch numbers of the products before the blinding operation. Rapid identification of product should also be possible in an emergency.

11.8 Recall, Return, Shipment and Destruction of IMP

Recall

Procedures for retrieving IMP's and documenting this retrieval should be agreed by the sponsor, in collaboration with the manufacturer or importer where different. The investigator and monitor need to understand their obligations under the retrieval procedure.

The Sponsor should ensure that the supplier of any reference or other medication to be used in a clinical trial has a system for communicating to the Sponsor the need to recall any product supplied.

Return

IMP's should be returned on agreed conditions defined by the sponsor, specified in approved written procedures. Returned IMP's should be clearly identified and stored in an appropriately controlled, dedicated area. Inventory records of the returned IMP's should be kept with the investigator.

Destruction of IMP

The Sponsor is responsible for the destruction of unused and/or returned IMP's. These products should therefore not be destroyed without prior written authorization by the Sponsor.

The delivered, used and recovered quantities of product should be recorded, reconciled and verified by or on behalf of the sponsor for each trial site and each trial period. Destruction of unused IMP's should be carried out for a given trial site or a given trial period only after any discrepancies have been investigated and satisfactorily explained and the reconciliation has been accepted.

Recording of destruction operations should be carried out in such a manner that all operations may be accounted for. The records should be kept by the Sponsor. When destruction of IMP's takes place a dated certificate of, or receipt for destruction, should be provided to the sponsor. These documents should clearly identify, or allow traceability to, the batches and/or patient numbers involved and the actual quantities destroyed.

Shipment of IMP

Shipping of IMP's should be conducted according to instructions given by or on behalf of the sponsor in the shipping order.

IMP should remain under the control of the sponsor until after completion of a two-step release procedure: certification by the qualified person; and fulfillment of the requirements of clinical trial.

The sponsor should ensure that these are consistent with the details actually considered by the qualified person. Both releases should be recorded and retained in the relevant trial files held by or on behalf of the sponsor.

De-coding arrangements should be available to the appropriate responsible personnel before IMP's are shipped to the investigator site.

A detailed inventory of the shipments made by the manufacturer or importer should be maintained.

Transfers of IMP's from one trial site to another should remain the exception. Such transfers should be covered by standard operating procedures. The product history while outside of the control of the manufacturer, through for example, trial monitoring reports and records of storage conditions at the original trial site should be reviewed as part of the assessment of the product's suitability for transfer and the advice of the qualified person should be sought. The product should be returned to the manufacturer, or another authorized manufacturer for re-labeling, if necessary, and certification by a qualified person. Records should be retained and full traceability ensured.

Products other than the test product, placebo or reference may be supplied to subjects participating in a trial. Such products may be used as support or escape medication for preventative, diagnostic or therapeutic reasons and/or needed to ensure that adequate medical care is provided for the subject. They may also be used in accordance with the protocol to induce a physiological response. These products do not fall within the definition of IMP's and may be supplied by the sponsor, or the investigator. The sponsor should ensure that they are in accordance with the notification/request for authorization to conduct the trial and that they are of appropriate quality for the purposes of the trial taking into account the source of the materials, whether or not they are the subject of a marketing authorization and whether they have been repackaged. The advice and involvement of a qualified person is recommended in this task.

Suggested Reading

1. Bert Spilker. Guide to clinical trials. Lipponcott Williams and Wilkins 2000
2. Richard A Guarino. New Drug Approval Process. Marcel Dekker Publication.
3. Steven Piontodosi. Clinical Trials-Methodology Perspective. Wiley publication.

Annexure

- IMP Accountability flow chart
- IMP Temperature Log
- IMP dispensing record.
- IMP disposition records (destruction record and return to Sponsor record)
- IMP-— label format

• **Example of a Label on Medication used in clinical research**
For Clinical Trials Use Only **Trial Name** For use in trial [give directions for use or as directed in patient information leaflet for the trial] Trial Protocol Non[XXXXXX] number. Name [of investigator and sponsor] Patient initials and identification number Date of supply Name and address of supplier

Investigational Medicinal Product Accountability Log (Study Code ____________________)

Center Number: ________Subject Number: ___________Patient Initials: _________ Randomization Number: _______

Visit Number	Medication	IMP DISPENSED						IMP RETURNED		
		Number of Pills/bottles Dispensed	Date of Dispense (mm-dd-yy)	Dispensed by Investigator / Pharmacist (Initials)	Date returned by patient (mm-dd-yy)	Number of days since last visit	Number of Pills/bottles returned	% Compliance (#dispensed-) #returned × 100 (# of days X #) no of pills prescribed/ day)	Completed by Investigator/ Pharmacist (Intials)	Comments

Investigational Medicinal Product (IMP) Accountability

Flow Chart

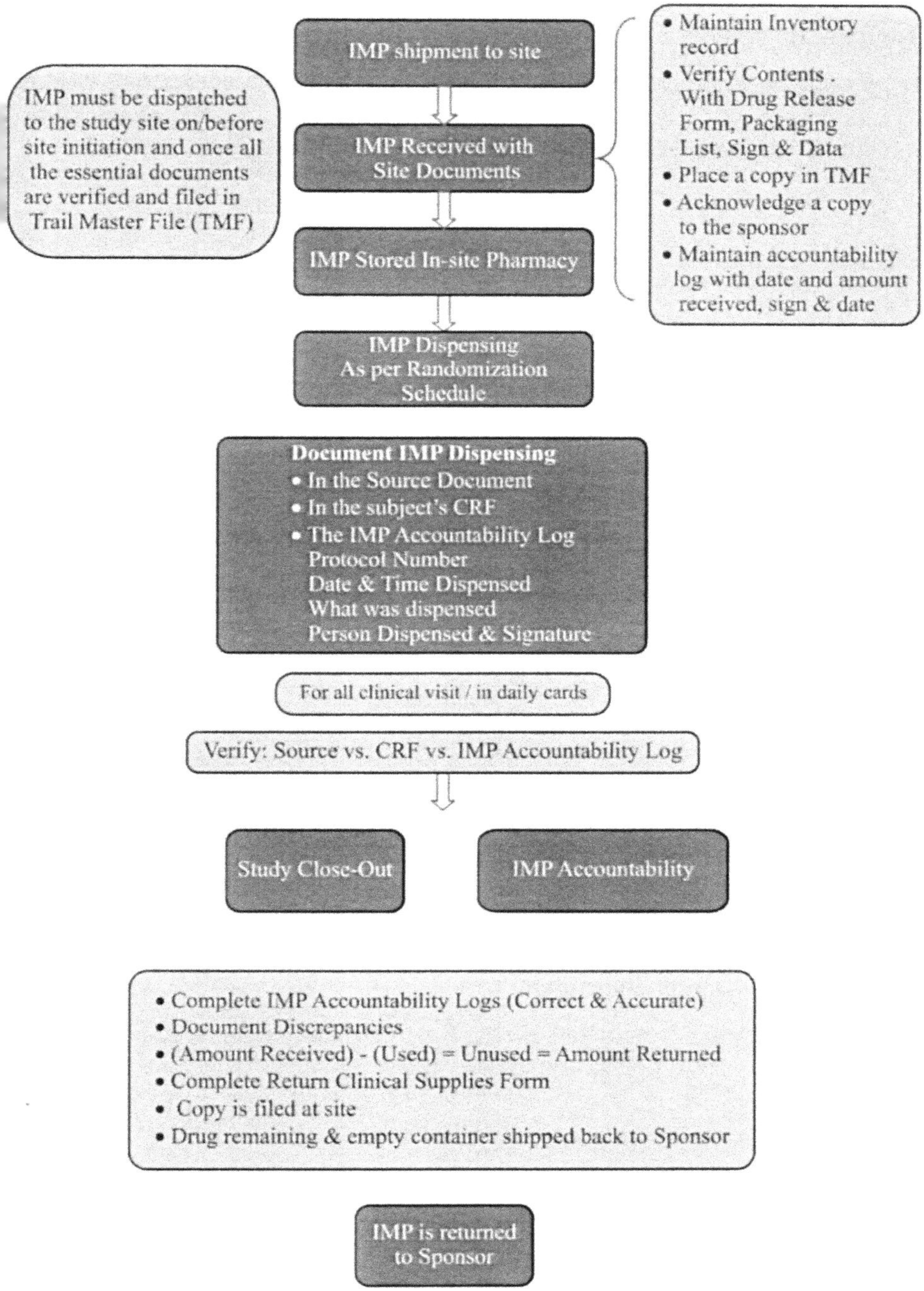

Temperature Log

In accordance with regulatory requirements, please record temperature weekly at a minimum (or daily if this is a standard practice at your site). A maximum-minimum thermometer is required.

The study medication storage temperature is between __°C - __°C

Date	Temperature (°C)		Initials	Comments
dd/mm/yyyy	Minimum	Maximum		

Please contact Field monitor and Local Trial leader if the temperature exceeds the maximum allowed temperature.

When this form is completed, please file in the study regulatory binder

Investigational Medicinal Product Destruction Form

IMP accountability completed by monitor/CRA on: ______________

(dd/mm/yyyy)

Monitor/CRA's Signature : __________________ Date: _______________

Site Staff signature : __________________Date: _______________

IMP to be Destroyed

Test Material Identification	Vial/Bottle/Package count
Protocol No: ______________ Batch No: ______________ Expiration Date: ___________	No. of used vials/ bottles/ packages: ______________ + No. of unused vials/ bottles/ packages: ______________ Total No. of used vials/ bottles/ packages: ______________ Date of Destruction: ___________

Note: Test material destruction must be approved and witnessed by the monitor/ CRA.

After completion please file it in the regulatory binder.

Investigational Medicinal Product Return /Transmittal Form

Project No:	Investigator Name:
Trial Code :	Sender
Country:	Date of Despatch
Centre No:	Method of Despatch

Name of the Field monitor/ Contact:

Details of Despatch

Material code	Sub. No. / Rand. code	Visit nos. / (or bulk)	Quantity e.g. 12	Units e.g. Capsules, vials, bottles etc	Other relevant information	Recipient check-Initials/ date

(Material Code: 1- New Trial medication, 2- Code break cards, 3- Treatment allocation cards, 4- Returned Trial medication from trial subjects, 5- Unused, undispensed trial medication, 6- Others (define)

Sender Name: Sender Title: Signature: Date :	Recipient Name: Recipient Title: Signature: Date :

CHAPTER 12

Multi-center Clinical Trials

T. Ramesh Kumar Rao

12.1 Introduction

A multi-center trial is a collective effort involving more than one independent center in enrolling and following study participants. A dramatic increase in the number of multi-center trials has been seen in the last two to three decades. The size of these trials varies depending on the requirements of the study. Though these studies are more difficult, expensive and bring less professional reward to the investigators, they are carried out because single centers cannot enroll enough participants.

By definition, a multi-center clinical trial is a trial conducted simultaneously by more than one investigator working in more than one institution but utilizing one common protocol. The aim of a multi-center trial is to pool the data collected and analyze them together.

A multi-center trial may have two or more participating institutions recruiting a relatively small number of patients over a few months to a trial having dozens of institutions recruiting hundreds of patients over a period of several years. The problems of co-ordination and standardization will be similar, regardless of the size of the trial, but more number of participants will make the trial more complex and any problems arising will be more difficult to overcome.

Anyone responsible for organizing and conducting a multi-center study should have a full understanding of the complexity of the undertaking. The problems for conducting such trials mostly originate from inadequate and unclear communication between the participating investigators, all of whom must agree to follow a common study protocol.

12.2 Advantages and Disadvantages

The advantages of multi-center clinical trials are

1. More rapid patient recruitment within a reasonable time.
2. More complex protocols may be able to be conducted because of additional resources utilized for certain large trials.
3. Less opportunity for one person's bias to influence the design or conduct of the trial.
4. Greater likelihood of high standards being attained for the data processing and analysis to be conducted.
5. Assures a more representative sample of the target population to be enrolled in the study.
6. Enables investigators with similar interests and skills to work together on a common platform.

The disadvantages of multi-center clinical trials are:

1. Administrative arrangements and management details are more complex.
2. Costs are usually greater for the clinical trial than if the same total number of patients were studied at a single site.
3. Statistical data analyses would be stronger from a single site.
4. Some ethics committees / IRBs may insist on changes to the protocol that may create major delays or are unacceptable to the sponsor or other Ethics Committees / IRBs.
5. Individual investigators in large multi-center trials receive little recognition through the publication(s) of results.

12.3 A Common Protocol

It is essential that the protocol designs are kept relatively simple and are the same at all the centers. If the data from several centers is to be pooled and analyzed together, it is obvious that they must have obtained the data by using similar methodology or the same protocol. A certain degree of "inter-center" variability may be taken into account in the analysis, but this variability should reflect no more than sampling fluctuations. Initial standardization must be done regarding patient recruitment, methods of assessing the results and techniques used for conducting biological tests.

Members from all the participating institutions should meet during the planning stage to discuss the protocol so as to ensure that all the terms are clearly and similarly understood by each of the members. In each institution, one should have a detailed discussion with the entire staff that is to be involved in the trial. Different interpretations of the protocol from one center to another may result in a lack of homogeneity. Hence, a protocol intended for a multi-center trial should be much more detailed than the one which is to be used by a single investigator and it should at the same time try to see all the questions that may arise during the conduct of a trial.

12.4 A Coordinating Centre

A coordinating centre consists of specialists and this is a requisite in all the multi-center trials. The members of this group must meet frequently and they should have sufficient time to direct and control the way a trial is set up and conducted. This includes careful planning of the initiation, designing and conduct of the trial. This centre is also responsible for implementing the randomization process, for carrying out day-to-day trial activities and for collecting, monitoring, editing and analyzing the data.

12.4.1 Composition of A Coordinating Centre

The composition of the coordinating group will vary depending on the nature of the trial but usually it includes one or two clinicians who are not doing the trial themselves, a statistician, a clinical pharmacologist, a representative of the sponsor and one or several specialists for the special examinations done in the trial. Its staff must have expertise in areas such as biostatistics, computer technology, epidemiology, medicine and management to respond expeditiously to daily problems that arise in a trial. The coordinating center should have adequate secretarial facilities, data processing equipment and should be readily contactable by telephone. The communication problems should be minimized through all possible techniques. Both formal and informal systems of communications should be established to include consideration of all aspects and also the organizational levels of the trial. In all, the staff at a coordinating center must be experienced, capable, responsive and dedicated to handle the workload in a timely fashion. A trial can succeed despite the inadequate performance of one or two clinical centers, but a poorly performing coordinating center can affect the success of a multi-center trial.

A coordinating center should have integrity. Any conflict of interest can damage the trial. This is one of the reasons for the sponsors to sometimes have outside organizations not involved in the trial as coordinating centers. Since the personnel in coordination center control the data and analyze it, they should have no overriding interest in the outcome of the trial.

Additional centers are often needed to perform specialized activities, such as conducting key laboratory tests, reading angiograms or x-rays, reading pathology slides and distributing study drugs. It is advisable not to permit a clinical center to perform these services. If a specialized center and a clinical center are in the same institution, each should have separate staff. Otherwise, unblinding and hence bias could result. Centers selected to perform specialized services need expertise in their particular fields. The center should have the capacity to handle such large workloads and at the same time not compromise on the research level quality expected of them.

12.4.2 Functions of a Coordinating Centre

The role of a coordinating center may vary from one trial to another but it generally involves the following functions:

- Designing and planning a protocol.
- Discussing the protocol with all the interested parties.
- Organizing pre-trial meetings.
- Writing up the final protocol.
- Motivating the investigators once the initial period of enthusiasm is over.
- Preparing or controlling the randomization procedure,
- The packaging of drugs used in the trial and the record forms used.
- Informing the investigators about how the trial is progressing, the adverse events that may have occurred, the difficulties encountered or the errors committed. Organizing meetings during the trial, preferably at different centers, to keep the participants abreast with what is being done in centers other than their own.
- Centralizing the data on patients from all centers, thus making it possible to perform the necessary verifications of data collection forms and to prepare interim reports and statistical analysis.
- Checking to ensure that the protocol is being properly followed and that there are no major violations with regard to selection criteria, treatment administered and dates of evaluation.

  Checking individual patient report forms to ensure that they have been filled appropriately. Should an error, omission or unlikely entry be detected, the correct information should be requested from the investigator as quickly as possible, while there is still time for him to re-examine the patient, check the result of a test or recall an event. Use of computers has become mandatory in all multi-center trials.

- Rejecting inadequate cases for which completion of the report form is unsatisfactory or for which there has been a major protocol violation.
- In extreme cases a participating center may have to be excluded if it becomes apparent that the data from this center is consistently unreliable.
- Replacing a center that is unable or unwilling to continue participating in a trial. If replacement is essential, the choice of another center should not be made lightly, since the new center will have to accept the protocol as it stands and all its constraints without having participated in its design and set up

12.5 Comparability of Participating institutions

The data collected from several centers cannot be analyzed together unless the centers performing the trial are comparable. In selecting the sites to include in a multi-center clinical trial one should make an attempt to include institutions that are comparable from the viewpoint of equipment, staff and types of patients recruited. It is therefore mandatory to conduct a survey of each organization likely to participate in the trial and observe the functioning in each center before commencing the trial.

12.5.1 Equipment

If the laboratory test is to be used as an assessment criterion in a multi-center trial, the sponsor should ascertain himself that the test can be performed under similar conditions in all the centers and that the results are equally reliable in each center. Samples may otherwise be sent to a single coordinating laboratory for testing to minimize the variability in the laboratory results. All equipments for measuring outcome criteria must be available at the centers and they must be standardized before the commencement of the trial.

12.5.2 Staff

The number of staff members, their availability, having the time and experience to conduct the clinical trial should be similar from one center to another. It is important that all the staff members understand the protocol definitions and how to complete the forms and perform the tests. Differences in performance among centers and between the individuals in a single center are unavoidable. They can, however be minimized by proper training, certification procedures, retesting and when necessary, retraining the staff. Only after the steering committee is satisfied that the staff are capable of performing necessary procedures should a clinical center be allowed to begin enrolling the participants into the trial.

12.5.3 Recruitment of Patients

The number of patients recruited in each center should not be too dissimilar and the sponsor should see that the inclusion and exclusion criteria are being followed to avoid the patient population from being heterogeneous with regard to demographic, pathologic and other characteristics. When centers recruit only a few cases the data quality may become mediocre if the data of this center is pooled for analysis.

12.5.4 Timing

If the trial participants are inpatients, there should not be major differences in the timing when the patients take their drugs, have their tests conducted or eat their meals.

12.6 Committees in a Multi-center Trial

In large multi-center trials, several committees might be required, having their areas of responsibility clearly established. They are the data monitoring committee, steering committee and investigators committee.

Data Monitoring Committee

The data monitoring committee should be independent of the investigators and the sponsor of the trial. It is responsible for periodically monitoring baseline, toxicity and response-variable data and at the same time evaluating the center's performance also. The independence of this group is important due to the recent concerns about clinical trial integrity. This committee reports to the sponsor of the study. The coordinating center should present tabulated and graphic data with appropriate analyses to the data monitoring committee. The committee is responsible for recommending to the sponsor of early trial termination in case of unanticipated toxicity, greater than expected benefit or high likelihood of indifferent results. The committee consists of members knowledgeable in the field of clinical trials methodology, biostatistics and an advocate may also be a part of this group. The committee is responsible for participant safety, more so, in double blind trials, since the individual investigators are unaware of the group assignments.

Steering Committee

In large multi-center studies, the steering committee, also called the executive committee, is made up of subset of investigators participating in the trial who are responsible for providing scientific direction for the study at the operational level. Subcommittees are formed to consider issues such as adherence to the protocol, quality control and publication policies. These subcommittees then report back to the steering committee. The issues requiring immediate decisions are addressed by this committee. It is important, however, that all major questions should be discussed with the investigators of all the participating centers.

Investigators Committee

All the investigators of the participating centers represent this committee. The principal investigator of each center is a voting member of this committee. The purpose of meetings of this committee is to keep all the investigators informed about the study progress of the study and it also provides an opportunity for staff training and education.

Conducting Additional Tests at one Site of A Multi-center Trial

In many multi-center trials one or more of the investigators often would like to conduct an additional test(s) or evaluation(s) as part of the trial. The main thing to consider before allowing or rejecting any of these studies is whether they will affect the main clinical trial in terms of patient cooperation, enrollment or compliance. Also, whether these additional studies will interact with tests already being performed and therefore, affecting the interpretation of the data? Any studies in which these issues could arise should be disallowed, even if an investigator feels they are important. The rule to remember is that the primary trial is the most important and the investigators should not be allowed to affect the results of the primary clinical trial by conducting additional studies.

If the sponsor accepts the additional test(s) or evaluation(s) then it should go as a protocol amendment and the investigator's ethical committee should approve this additional test(s) or evaluation(s). The performing of this additional test(s) or evaluation(s) may be limited to one site only or made optional for the other sites as well.

12.7 Statistical Issues

Should Data from Each Site be Able to Stand Alone?

The reason for conducting a multi-center trial is on the assumption that any single center will not be able to enroll an adequate number of patients. However, the data obtained from various sites can be combined only if the data is found to be compatible. The compatibility of data at different centers cannot be forecast in advance for multi-center trials as this depends on the enrollment of adequate number of patients. If this is done, then the data from any site may be separated from the other sets of data. Each site may then be analyzed separately and still retain reasonable statistical power.

A Wise Precaution: Pilot Studies

Once the centers participating in a multi-center trial have been identified and the protocol finalized, it may be a good idea to ask each center to do a pilot study with a limited number of patients to provide the following information about:

1. Expected patient accrual in each center.
2. The feasibility of the study (the protocol may not be realistic)
3. Regarding the quality of the data collected, which may require discussion between coordinating center and all participating institutions about errors or omissions which will have to be carefully watched for.

4. The homogeneity between centers regarding the patients enrolled and th manner in which data is being collected.
5. Patient acceptability of the treatments and side effects due to th treatments.
6. The variability in the results, which may make it possible to calculate th power of the tests used to compare the groups of a given size.
7. The results of a pilot study may sometimes reveal factors that have strong bearing on the outcome of a trial and such factors may have to b dealt with stratification.

After the completion of the pilot study, the question is whether to include th results of the study in the final analysis. This is acceptable if no changes a made in the protocol following the pilot study. In fact, a protocol seldom remair unchanged after a pilot study since the purpose of a pilot study is to detect an rectify the weak points in a protocol.

12.8 Analysis of Results

The following are to be considered while analyzing the results:

1. If the differences between treatments vary outside random fluctuatio in different centers, the cause should be looked for:
   (a) The generalization of results should be restricted for certai categories of patients if there are differences in patient selectio among the centers.
   (b) Bad adherence to protocol or insufficient motivation of investigato should beconsidered if one sees abnormal frequency of losses t follow-up or non-compliers in one or several centers.
   (c) If there are a small number of outlier cases in a center it ma decrease the credibility of the data, but it does not require th exclusion of the center from its data being analyzed.
2. The number of patients recruited by each investigator should be considere Although it is not ideal, differences of sample sizes are frequent betwee centers. If a particular center has a poor recruitment, its data need not b pooled for analysis.

Controversial Outcomes

Careful planning, conduct and analysis of a large multi-center trial does n ensure that the methodologies used, data gathered or analyses performed wi be accepted by the medical community or by regulatory authorities. In the la two decades there has been strong criticism for some of the large multi-cent trials (e.g., University Group Diabetes Program [UGDP] Trial). There is r simple solution to the dilemma of spending enormous sums of money, energ time, patient involvement and other resources on large multi-center trials and yet having the basic approaches or interpretations of the data being strongly criticized.

Periodic Meeting and Final Reports

Interim or annual meetings of the investigators, monitors and other trial personnel should be conducted during the conduct of a large multi-center clinical trial. Meeting of the committees formed should also be held on a periodic basis.

Publication, presentation and authorship policies should be determined in advance. Authorship becomes a critical issue when there are multiple investigators, many of whose academic careers depend on publication.

The results of a multi-center clinical trial should be published in a widely read journal so as to reach a wider readership. In addition to this, Methodological papers should be published, which describe experiences in specific aspects of a clinical trial (e.g., recruitment, randomization, storage of medicine, follow-up visits, dropouts) and also spell out the lessons learned.

12.9 Multinational Clinical Trials

A multinational clinical trial is a single trial conducted in two or more countries and the data is combined. All multinational trials are multi-center trials, but not all multi-center trials are multinational. The major reasons for conducting a multinational trial are:

1. Gaining more rapid patient enrollment.
2. Expediting the development of a new medicine.
3. Conducting a trial that otherwise might be impossible.
4. Gaining experience and being able to compare patients of different nationalities in the same trial.

The most important principle is that multinational clinical trials should not be conducted unless necessary. The following procedures should be adhered to when such a trial is conducted:

1. Data should be examined for national differences.
2. One or two well-defined simple endpoints should be used. These should be the most well-validated measures.
3. High scientific and clinical standards should be used.
4. Monitoring should be of a high standard and conducted at all sites in all countries.
5. The clinical trial should be within a therapeutic area in which national differences among the subjects are not expected to affect the diagnosis or treatment significantly.

6. The clinical trial should not be part of Phase I and Phase II (ideally) medicine development.
7. The clinical trial should not be a quality of life trial because large cultural differences among the subjects will make the interpretation of the data impossible.
8. It is important to hold a pretrial meeting, where all the aspects of the protocol and clinical trial are discussed.
9. In most cases two or more national multi-center clinical trials will provide stronger evidence of a medicine's efficacy than two or more equally well conducted multinational clinical trials.
10. Careful attention has to paid to the data collection forms so as to ensure that comparable data are collected. Translation and retranslation to the original language has to be done.

The specific pointers in designing, conducting and interpreting multinational clinical trials are:

1. Ensure that input is received from at least one physician consultant in each country participating in the clinical trial both prior to and during the protocol's preparation.
2. Ensure that as many physicians as possible review the protocol prior to submitting it for Ethics Committees review.
3. Include those sites in the countries where clinical practice and culture are generally similar.
4. Ensure that an appropriate audit of the data is made by the sponsor or regulatory authority.
5. Consider a pilot trial to show that the larger trial is feasible and also to identify potential problem areas that will need to be addressed.
6. The single group of monitors visiting all the sites should be trained appropriately in manners and behavior as they will be meeting investigators from different cultures as they are from different countries.
7. Countries that differ markedly in medical practice, culture, regulations or practical aspects (e.g., Japan) should not be part of a multinational clinical trial.

General Comments

Even if investigators think that they have identified all the potential difficulties and have taken care to prevent them, new problems will always arise. This is particularly true of multi-center trials because of their size, complexity and large number of investigators with diverse backgrounds of interests. To minimize problems, adequate communication must be stressed. It is the responsibility of the coordinating center to keep in frequent contact with all the centers. The principal investigators need to maintain contact with all the participating centers and committees that are closely monitoring the trial.

It is always difficult to get groups of clinicians to work together using a common problem.

Cost is always a concern in multi-center trials. These studies are generally expensive because of their complexity, size and elaborate committee structures. Expense can be minimized by asking only pertinent questions on forms, reducing the number of laboratory tests and performing only necessary quality monitoring. To accomplish these goals it demands constant attention especially during the planning phase.

The purpose of every procedure and item on the study forms should be clearly defined in advance and test hypothesis formulated, if possible. Thought should be given to sample size and power calculations.

It is always a good policy to restrict additions to the basic study protocol. If additions are allowed the trial may get out of control.

Suggested Reading

1. John I Gallin. Principles and practise of clinical research. Elsevier Publication.
2. Shein Chung Chow. Design and analysis of clinical trials. (Concepts and methodologies). Wiley Publication

CHAPTER 13

Good Clinical Research Practice

T. Ramesh Kumar Rao

13.1 Introduction

Clinical research is necessary to establish the safety and effectiveness of specific health and medical products. The knowledge of safety and efficacy of specific products and treatments has come from randomized controlled clinical trials that are designed to answer important scientific and health care questions. These randomized controlled clinical trials form the foundation for "evidence based medicine", but such research can be relied upon only if it is conducted according to principles and standards referred to as "Good Clinical Research Practice" (GCP).

These guidelines for GCP are intended to assist national regulatory authorities, sponsors, investigators and ethics committees in implementing GCP for industry, government and/or institution sponsored or investigator initiated clinical research. These guidelines were issued subsequent to the 1995 International Conference on Harmonization Good Clinical Practice (ICH-GCP). The ICH – GCP guidelines are a reference tool for understanding and implementing GCP by:

- Describing the clinical research process as it relates to health and medical products, and identifying and explaining each of the activities that are common to most trials to the parties who are usually responsible for carrying them out.

- Linking each of these processes to one or more principles of GCP.
- Explaining each GCP principle and providing guidance on how each principle is routinely applied and implemented.

13.2 Definition of GCP

Good Clinical Research Practice (GCP) is a process that incorporates established ethical and scientific quality standards for the design, conduct, recording and reporting of clinical research involving the participation of human subjects.

Compliance with GCP assures the public that the rights, safety and well being of the research subjects are protected and respected according to the principles laid down in the Declaration of Helsinki and other internationally recognized guidelines. It also ensures the integrity of the clinical research data. The conduct of clinical research is complex as it requires involvement of a number of individuals with a variety of expertise, performing their allotted tasks skillfully and efficiently.

13.3 Overview of the Clinical Research Process

The key activities involved in the conduct of a clinical trial are enumerated below:

1. ***Development of the trial protocol:*** The GCP requires that the clinical trials should be described in a clear, detailed procedure. The sponsor, often in consultation with one or more clinical investigators, designs the study protocol. Clinical investigators may also design and initiate clinical studies. The sponsor and clinical investigator(s) should be aware of the national / local regulations regarding the designing, initiating and conduct of the study.
2. ***Development of standard operating procedures (SOPs):*** All parties who oversee, conduct or support clinical research (i.e., sponsors, clinical investigators, Independent Ethics Committees / Institutional Review Boards [IECs / IRBs], monitors, and contract research organizations [CROs]) should develop and follow written SOPs defining the responsibilities, records to be maintained and methods to be used for study related activities.
3. ***Development of support systems and tools:*** Appropriate support systems and tools will facilitate the conduct of the study and collection of data as required by the protocol. The tools in trial related information documents include investigator's brochure, case report forms (CRFs), checklists, study flow sheets, and drug accountability logs. Those not related to trial-related information documents are computer hardware and software, electronic patient diaries and specialized equipment.

   The sponsor is usually responsible for developing, maintaining, modifying and ensuring the availability of support systems and tools for conducting the trial and also for collecting and reporting the required data.

4. ***Generation and approval of trial-related documents:*** The development of trial-related documents facilitates the conduct of the study, collection as well as reporting of study-related data and analysis of study results. The sponsor generally develops, designs and provides various standardized forms and checklists to assist the clinical investigator and his / her staff in capturing and reporting the data required by the protocol.
5. ***Selection of trial sites, qualified, trained and experienced investigators and study personnel:*** The sponsor should review the requirements of the study protocol to determine the type of expertise required and identify clinical investigators with necessary medical expertise to conduct the study. The clinical investigator should also have knowledge, training and experience in the conduct of clinical trials and human subject protection. The clinical investigator should have sufficient resources and appropriately trained staff to conduct the study.
6. ***Ethics committee review and approval of the protocol:*** The GCP requires that the studies must be reviewed and receive approval / favourable opinion from an independent IEC / IRB prior to enrollment of study subjects. The investigator is responsible for obtaining this approval, which is then provided to the sponsor.
7. ***Review by regulatory authorities:*** The GCP requires that studies undergo review by regulatory authorityies for using the investigational product or intervention on human subjects and also to assure that the study is appropriately designed to meet its stated objectives according to the national regulations. [Note: Some countries may not have systems in place for reviewing research or may depend on external review, while some may have additional requirements for the review and approval of trial sites and / or investigators.

   The sponsor is generally responsible for ensuring the applicable regulatory authorization before the commencement of the study.
8. ***Enrollment of subjects into the study: recruitment, eligibility and informed consent:*** The clinical investigator is responsible for recruiting subjects. He should ensure only eligible subjects are enrolled into the study. He should obtain and document the informed consent of each subject. The GCP requires that the informed consent must be obtained from each subject prior to enrollment in the study or performing any study related procedures.
9. ***The investigational product(s): quality, handling and accounting:*** Quality of the investigational product is assured by compliance with Good Manufacturing Practices (GMPs) and handling and storing the product according to the manufacturing specifications and the study protocol. GCP requires that the sponsor control the access to the investigational product and also maintain documents regarding the quantities produced, to whom

the product is shipped and handed over and also the return or destruction of any unused supplies. GCP also requires the investigators to maintain a record of receipt, administration and disposition of the investigational product.

10. ***Trial data acquisition: conducting the trial:*** Research should be conducted according to the approved protocol and applicable regulatory requirements. The clinical investigator should also document each trial-related activity in the study records which will verifiy if the study was protocol compliant.

11. ***Safety management and reporting:*** All clinical trials must be managed for safety. Although all parties who oversee or conduct clinical research have a role / responsibility for safety of the study subjects, the clinical investigator has primary responsibility for alerting the sponsor and IEC/ IRB to serious / life threatening unanticipated events (SAEs) observed during the course of the research. Then it is the sponsor's responsibility for reporting the SAEs to regulatory authorities and other investigators if it is a multi-centric trial. The sponsor may constitute a data and safety monitoring board (DSMB) to assist him in overall safety management.

12. ***Monitoring the trial:*** Sponsors usually perform site monitoring of a clinical trial to assure high quality trial conduct, directly or utilizing the services of an outside organization (e.g., contract research organization [CRO].

    The monitors review individual case histories to verify adherence to the protocol and also review data entry and quality control procedures. It is also the duty of the monitor to verify if the investigator is adhering to GCP. In blinded studies, these monitors also remain blinded to study arm assignment.

13. ***Managing trial data:*** The GCP requires the appropriate management of the clinical trial data so as to assure that the data collected is complete, reliable and processed correctly. This enables the preservation of the data integrity. Data management includes all processes and procedures for collecting, handling, manipulating, analyzing and storing / archiving of data from start to completion of the study.

    It is the sponsor's responsibility to develop appropriate data management systems. It is the responsibility of both the sponsor and the investigator for implementing such systems to ensure that the integrity of the trial data is preserved.

14. ***Quality assurance of trial performance and data:*** Quality assurance (QA) verifies through systematic independent audits whether the existing quality control systems (e.g., study monitoring) are working and effective. The quality assurance audits may be performed during the course of the clinical trial and/or upon trial completion. It is the sponsor's responsibility to establish the quality systems and conduct quality assurance audits.

15. ***Reporting the trial:*** The results of each controlled study involving an investigational product should be summarized and described in a clinical study report containing clinical data, statistical descriptions, presentations and analyses. The report should be complete, timely, well-organized, free from ambiguity and easy to review. The sponsor is responsible for preparing clinical study reports. Where permitted, abbreviated or less detailed reports may be acceptable for uncontrolled or aborted studies.

13.4 Principles of GCP

1. A clinical trial should be conducted in accordance with ethical principles (as in the Helsinki document) and be consistent with the GCP and all regulatory requirements.
2. Clinical trials must be scientifically sound and described in a clear, detailed protocol.
3. The pre-clinical / non-clinical (animal) data found in the Investigator's Brochure must be adequate to support the proposed clinical trials.
4. The rights, safety and well being of the trial subjects must remain the top priority.
5. A trial must be conducted exactly as stated in the protocol that has received prior oversight committee (IRB, IEC, REB) approval.
6. A freely given informed consent must be obtained from every subject prior to the start of the study.
7. Risks should be weighed against anticipated benefits and a study continued only if the anticipated benefits justify the risks to the subject.
8. The medical care given to, and medical decisions made on behalf of subjects, should always be the responsibility of a trained and qualified physician.
9. Each individual involved in conducting a trial should be qualified by education, training and experience.
10. All clinical trial information (raw data) must be recorded, handled and stored in a way that ensures that it is accurately reported, interpreted and verified. You must be absolutely certain that there is a clear data trail by documenting each and every decision at the time it is made.
11. The absolute confidentiality of study participants/subjects and their records must be protected so that the identity of subjects remains unknown by all except the PI (and her/his designates).
12. Investigational products must be used only as per the protocol and must be manufactured, handled and stored in accordance with Good Manufacturing Practice (GMP).

13. Systems/procedures that assure the quality of every aspect of the trial are to be implemented in a standard repeatable way such as a clearly written, understood and authorized Standard Operating Procedures known in the industry as the "SOP's".

Principle 1: Ethical Conduct

A clinical trial should be conducted in accordance with ethical principles (as in the Helsinki document) and be consistent with the GCP and all regulatory requirements.

Ethical principles have been established by many national and international bodies including:

(a) The World Medical Association Declaration of Helsinki

(b) The Council for International Organizations of Medical Sciences (CIOMS) International Ethical Guidelines for Biomedical Research Involving Human Subjects.

Application

Principle 1 is applied through:

1. Design and approval of the protocol
2. Informed consent
3. Scientific and ethical review
4. A favourable risk/benefit assessment
5. Fair and transparent procedures and outcomes in the selection of research subjects
6. Compliance with national and international laws, regulations and standards

Implementation

The basic ethical principles of biomedical research are reflected in all GCP principles and processes, impacting the role and responsibilities of each party within the GCP. This includes the investigator(s) and site staff, the sponsor and sponsor's staff (including monitors and auditors), the ethics committee(s), the regulatory authority(ies) and the individual research subjects.

Principle 2: Protocol

Clinical trials must be scientifically sound and described in a clear, detailed protocol.

"The experiment should be such as to yield fruitful results...unprocurable by other methods or means of study and not random and unnecessary in nature."(The Nuremburg Code)

"The design and performance of each experimental procedure involving human subjects should be clearly formulated in an experimental protocol". (Declaration of Helsinki)

Application

Principle 2 is applied through development of a clear, detailed, scientifically justified and ethically sound protocol that:

(a) Complies with requirements established by national and local laws and regulations

(b) Undergoes scientific and ethical review prior to implementation

Implementation

The sponsor is responsible for:

(a) Designing the clinical investigation

(b) Developing the study protocol, investigator's brochure, related materials to describe the procedures that will be followed, study endpoints, data collection and any other study requirements.

(c) Ensuring the protocol complies with applicable national and local laws and regulations.

Investigators may be consulted by the sponsor during the protocol development and in some cases the investigator may personally contribute to the design of the protocol. Investigators are responsible for familiarizing themselves with the study protocol, investigator's brochure and any other study related matters to ensure that they are able to carry out the study in compliance with the specifications of the protocol.

IECs/IRBs are responsible for conducting ethical review of the study protocol. This also includes arranging for a scientific review or verifying that a competent body has determined that the research is scientifically sound.

Regulators bear responsibility for allowing a protocol to proceed in accordance with applicable laws and regulations. This may include prospective review of the protocol, the investigator's brochure and other relevant information. When the protocol or investigator's brochure is inaccurate or materially incomplete or when the protocol does not adequately provide for the protection of subject rights and safety or when the protocol is deficient in design to meet its stated objectives the regulatory authority may request for protocol modification or take action to disallow the protocol or proceed in accordance with applicable laws and regulations.

Principle 3: Risk Identification

The pre-clinical / non-clinical (animal) data found in the Investigator's Brochure must be adequate to support the proposed clinical trials.

"The experiment should be so designed and based on the results of animal experimentation and a knowledge of the natural history of the disease or other problems under study, that the anticipated results will justify the performance of the experiment."(The Nuremberg Code)

"Medical research involving human subjects must conform to generally accepted scientific principles, be based on a thorough knowledge of the scientific literature, other relevant sources of information, and on adequate laboratory and appropriate animal experimentation."(Declaration of Helsinki)

"The assessment of risks and benefits requires a careful arrayal of relevant data, including, in some cases, alternative ways of obtaining the benefits sought in the research...[T]he assessment presents both an opportunity and a responsibility to gather systematic and comprehensive information about proposed research."(The Belmont Report)

Application

Principle 3 is applied through:

(a) A thorough search conducted on the available scientific information about the investigational product or procedures (including findings from tests in laboratory animals and any previous human experience).

(b) Developing the investigator's brochure, the study protocol and the informed consent document to adequately reflect the available scientific information on foreseeable risks and anticipated benefits.

Implementation

The responsibility for implementing this principle is shared by the sponsor, investigators, IECs/IRBs and regulators.

The sponsor conducts literature review to ensure that sufficient information is available to support the proposed clinical trial in the population to be studied and there is sufficient safety and efficacy data to support human exposure to the product. The sponsor may have to conduct pre-clinical studies to ensure that it has sufficient safety and efficacy data to support human exposure. The sponsor summarizes the available information about the procedure/product in the investigator's brochure and designs the study protocol accordingly. It is the sponsor's responsibility to develop a comprehensive, accurate and complete investigator's brochure and this is the means for communicating vital safety and scientific information to the investigator who in turn communicates it to the IEC/IRB.

Review of the protocol, investigator's brochure and other relevant information enables the IEC/IRB to:

(a) Determine whether the benefits outweigh the risks.

(b) Understand the study procedures or other steps that will be taken to minimize risks.

(c) Ensure that the informed consent document accurately states the potential risks and benefits in a way that it will facilitate comprehension by all study related subjects, with particular attention to vulnerable groups.

Investigators must be knowledgeable of the protocol, investigator's brochure and other relevant information regarding potential risks and benefits. They must be able to adequately, accurately and objectively identify the potential risks and benefits to subjects. The investigators may have to do additional literature search to gather information beyond that provided by the sponsor. They have to be thoroughly familiar with the appropriate use of the trial product(s)/procedures and should take necessary steps to remain aware of all relevant new data that becomes available during the course of the clinical trial.

Regulators bear the responsibility for allowing a protocol to proceed in accordance with existing national laws/regulations or internationally accepted standards. This includes a prospective review of the protocol, investigators brochure and other relevant information to ensure that risk(s) and benefit(s) are accurately identified and justify the protocol to proceed. Adopted national standards should also address additional national or regional racial, cultural or religious standards/issues which are not covered by international standards. In accordance with the national/local laws and regulations, regulators may establish standards for the conduct of non-clinical studies, review non-clinical and clinical data submitted in support of research permits or marketing applications and/or inspect facilities that conduct non-clinical and clinical studies.

Principle 4: Benefit – Risk Assessment

The rights, safety and well being of the trial subjects must remain the top priority.

"The degree of risk to be taken should never exceed that determined by the humanitarian importance of the problem to be solved by the experiment."(The Nuremberg Code)

"Every medical research project involving human subjects should be preceded by careful assessment of predictable risks and burdens in comparison with foreseeable benefits to the subject or to others. This does not preclude the participation of healthy volunteers in medical research." (Declaration of Helsinki)

"For all biomedical research involving human subjects, the investigator must ensure that potential benefits and risks are reasonably balanced and risks are minimized." (CIOMS, International Ethical Guidelines, Guideline 8)

"It is commonly said that benefits and risks must be 'balanced' and shown to be 'in a favourable ratio' .... Thus, there should first be a determination of the validity of the presuppositions of the research; then the nature, probability and magnitude of risk should be distinguished with as much clarity as possible. The method of ascertaining risks should be explicit... It should also be determined whether... estimates of the probability of harm or benefits are reasonable as judged by known facts or other available studies." (The Belmont Report)

Application

Principle 4 is applied through appropriate study design and through ethical, scientific and where applicable, regulatory review of the study protocol prior to its initiation.

Implementation

The responsibility for implementing this principle is shared by sponsors, investigators, IEC(s)/IRB(s) and regulators.

The sponsor should design research studies where the risks to subjects will be minimal.

The investigator(s) should review the investigator's brochure and other relevant risk and benefit information in making a decision to conduct the study. The investigator is also responsible for providing adequate, accurate and objective information on risks and benefits during informed consent of study subjects.

The regulatory authority may require modification to a protocol as a condition to its proceeding and/or may suspend or terminate a protocol based on an unacceptable risk/benefit profile in accordance with applicable laws and regulations.

Principle 5: Review By Iec/Irb

A trial must be conducted exactly as stated in the protocol that has received prior oversight committee (IRB, IEC, REB) approval.

The "...protocol should be submitted for consideration, comment, guidance and where appropriate, approval to a specially appointed ethical review committee, which must be independent of the investigator, the sponsor or any other kind of undue influence. This independent committee should be in conformity with the laws and the regulations of the country in which the research experiment is performed..."(Declaration of Helsinki)

"Failure to submit a protocol to the committee should be considered a clear and serious violation of ethical standards" (CIOMS, International Ethical Guidelines, Commentary to guideline 2)

Application

Principle 5 is applied for protocol review by an IEC/IRB that is constituted and operating in accordance with GCP and applicable national/local laws and regulations.

Implementation

The responsibility for implementing this principle is shared by IEC(s)/IRB(s), investigators, sponsors and regulators.

A properly constituted and operational IEC/IRB will review the protocol (and/or any proposed amendments to the protocol) and provide the investigator with a written decision/opinion. The IEC/IRB written procedures should ensure that no subject be enrolled into a trial or any deviations or changes to the protocol are initiated before obtaining the IEC's /IRB's approval/favourable opinion.

The investigators should submit the study protocol to their IEC(s)/IRB(s) and it is their responsibility for obtaining an approval/favourable opinion prior to enrolling any subjects into the trial. Investigator's should not deviate or make any changes in the protocol without prior permission from the sponsor and prior review and documented approval/favourable opinion from the IEC(s)/IRB(s), except where it is necessary to eliminate an immediate hazard(s) to participating trial subjects.

The sponsor should develop the protocol, select qualified investigator(s)/institution(s) and conform that each of the investigator has had the study protocol reviewed by an IEC/IRB and received the IEC/IRB approval/favourable opinion.

In accordance with the applicable laws/regulations, regulators may inspect the investigator(s), sponsor(s) and/or IEC(s)/IRB(s) to ensure compliance with IEC/IRB review requirements. Regulators should also encourage the IEC(s)/IRB(s) to communicate with them directly on issues or concerns that they may encounter in their review of human trials.

Principle 6: Informed Consent

A freely given informed consent must be obtained from every subject prior to the start of the study.

"In particular, no one shall be subjected without his free consent to medical or scientific experimentation."(United Nations International Covenant on Civil and Political Rights)

"The subjects must be volunteers and informed participants in the research project."(Declaration of Helsinki)

'...[T]here is widespread agreement that the consent process can be analyzed as containing three elements: information, comprehension and voluntariness."(The Belmont Report)

"For all biomedical research involving humans, the investigator must obtain the voluntary informed consent of the prospective subject or, in the case of an individual who is not capable of giving informed consent, the permission of a legally authorized representative in accordance with the applicable law. Waiver of informed consent is to be regarded as uncommon and exceptional and must in all cases be approved by an ethical review committee."(CIOMS, International Ethical Guidelines, Guideline 4)

'Obtaining informed consent is a process that begins when initial contact is made with a prospective subject and continues throughout the course of the study. By informing the prospective subjects, by repetition and explanation, by answering their questions as they arise and by ensuring that each individual understands each procedure, investigators elicit their informed consent and in so doing manifest respect for their dignity and autonomy."(CIOMS, International Ethical Guidelines, Commentary on Guideline 4)

Application

Principle 6 is applied as a process of informing and ensuring comprehension by the study subjects (and/or their legally authorized representatives) about the research and then obtaining their consent, including written informed consent.

Implementation

The responsibility for implementing and overseeing the informed consent process is shared by the sponsors, clinical investigators, IEC(s)/IRB(s) and regulatory authorities.

IEC(s)/IRB(s) are responsible for:

(a) Reviewing the informed consent document and ensuring that it is accurate, complete and written in a language that will be understood by the potential study subjects and then translated into other languages that the study subjects are likely to understand.

(b) Requesting modifications to the informed consent document, if required.

(c) Observing the consent process and research is at their discretion.

Investigators are responsible for ensuring that:

(a) Staff responsible for obtaining informed consent should receive appropriate training, both in research ethics and in the requirements of the study protocol

(b) The IEC/IRB should review and approve the informed consent form and other written information that is to be used in the study prior to its use.

(c) Informed consent should be obtained from each of the study subjects or their representatives prior to involving the subject in any study related activities, including diagnostic or other tests that are administered solely for determining the subjects' eligibility to participate in the research.

Sponsors are responsible for monitoring the research at study sites and also ensure that sites are obtaining informed consent from all study subjects prior to their inclusion in the research study.

Regulators may inspect the various parties who conduct or oversee research to ensure that they are complying with applicable laws and regulations and enforcing non-compliance. Regulators may therefore inspect IEC(s)/IRB(s) to ensure that informed consent documents and procedures are appropriately reviewed, inspect the clinical investigators to determine whether informed consent was obtained prior to subjects' inclusion in the study and also inspect the sponsors to ascertain that they are appropriately monitoring the studies.

Principle 7: Continuing Review / Ongoing Benefit-Risk Assessment

Risks should be weighed against anticipated benefits and a study continued only if the anticipated benefits justify the risks to the subject.

"During the course of the experiment the scientist in charge must be prepared to terminate the experiment at any stage, if he has probable cause to believe, in the exercise of the good faith, superior skill and careful judgment required of him that a continuation of the experiment is likely to result in injury, disability or death to the experimental subject."(The Nuremburg Code)

"...The ethical review committee should conduct further reviews as necessary in the course of the research, including monitoring of its progress." (CIOMS, International Ethical Guidelines, Guideline 2)

"... The committee has the right to monitor ongoing trials..."(Declaration of Helsinki)

"...Clinical trial sponsors should develop a process to assess, evaluate and act on safety information during drug development on a continuous basis in order to ensure the earliest possible identification of safety concerns and to take appropriate risk minimization steps. Such steps can include modification of study protocols, to incorporate strategies to ensure that clinical trial participants are not exposed to undue risk."(Management of Safety Information from Clinical Trials, Report of CIOMS Working Group VI. Identification and Evaluation of Risk from Clinical Trial Data)

Application

Principle 7 is applied to develop and implement the process for evaluating risks and benefits of the research as additional information becomes available during the course of the study. This principle encompasses

1. Safety monitoring of the study by investigator(s) and sponsor (including use of a data and safety monitoring board [DSMB], where appropriate.
2. Reporting serious and unexpected adverse events or other unanticipated risks to the sponsor, IEC/IRB and regulators.
3. Review by the IEC/IRB of any unanticipated risks as they occur or at scheduled intervals appropriate to the degree of risk..
4. Reviewing the protocol, investigator's brochure, informed consent document as needed and suspending or terminating the studies if necessary to protect the rights and welfare of the study subjects.

Implementation

Sponsors, DSMBs (if applicable), IEC(s)/IRB(s) and regulators share the responsibility for ongoing safety evaluations of the investigational product(s).

The investigator reports unanticipated problems involving risks to subjects and provides periodic progress reports at intervals appropriate to the degree of risk to sponsors and IEC(s) / IRB(s) in accordance with the national/local laws and regulations. The investigator should provide adequate, accurate and objective information on risks and benefits during informed consent of study subjects and renew the consent of the subject to continue in the study, as appropriate.

The sponsor monitors the study and performs safety evaluations of the investigational product(s) by analyzing data received from the investigator(s) and the DSMB (if one has been appointed). The sponsor also assures expedited reporting of both serious and unexpected adverse events to the regulatory authorities. He also assures that the same is reported by the investigator(s) to the IEC(s) / IRB(s).

As the study progresses, the IEC(s)/IRB(s) conduct follow up reviews appropriate to the degree of risk, at least once a year. They review the investigator's progress reports to determine if the benefits still outweigh the risks.

The regulatory authority reviews the data submitted in research or marketing permits and may ask for a modification to the protocol as a condition for the trial to proceed or it may suspend or terminate a protocol based on an unacceptable benefit-risk profile in accordance with applicable laws and regulations.

Principle 8: Investigator Qualifications

The medical care given to, and medical decisions made on behalf of subjects, should always be the responsibility of a trained and qualified physician.

"The experiment should be conducted only by scientifically qualified persons. The highest degree of skill and care should be required through all stages of the experiment of those who conduct or engage in the experiment."(The Nuremberg Code)

"Medical research involving human subjects should be conducted only by scientifically qualified persons and under the supervision of a clinically competent medical person..."(Declaration of Helsinki)

Application

Principle 8 defines the sponsor's responsibility in the selection of qualified investigator(s) and also the responsibilities of the clinical investigator to the study subject. (See also GCP Principle 9, Staff Qualifications)

Implementation

The investigator is responsible for providing, or ensuring that study subjects have access to medical care for medical problems arising during their participation in the trial and for following the subjects' status until the problem is resolved.

"It is recommended that the investigator inform the subject's primary physician about the subject's participation in the trial if the subject has a primary physician and if the subject agrees to the primary physician being informed."(ICH E6, Section 4.3)

The primary responsibility for selecting qualified clinical investigators to conduct a study resides with the sponsor.

The IEC(s) / IRB(s) is responsible for ensuring that the rights and welfare of study subjects are protected. Consideration of investigator qualifications and experience and the adequacy of the site (including the supporting staff, available facilities and emergency procedures) by the IEC/IRB will ensure that the subjects have access to appropriate care for medical problems arising during the participation in the trial.

The national and / or local regulatory authorities have indirect responsibility related to clinical investigator qualifications.

Principle 9: Staff Qualifications

Each individual involved in conducting a trial should be qualified by education, training and experience.

GCP requires that the clinical investigator is appropriately qualified by education, training and experience to conduct the clinical trial. GCP also requires that each clinical investigator will have adequate resources available, including sufficient staff, who are also appropriately qualified by education, training and experience to assist him / her with the trial and ensure the safety of study subjects.

Application

Principle 9 is chiefly applied through the clinical investigator's selection of appropriate staff to assist him with the conduct of the study.

Implementation

It is the primary responsibility of the investigator to:

1. Select qualified staff to assist in the conduct of the investigation.
2. Ensure that the study staff receive appropriate training related to ethics and consent procedures and other requirements as per the study protocol.
3. Establish clear procedures for activities related to the conduct of the study.
4. Assign tasks to staff, based on their qualifications, experience and professional licenses.

5. Personally supervise the staff to ensure that they are diligently fulfilling their study related duties.

Although the investigator may delegate tasks to members of his/her staff, it is the investigator who is responsible for the study and he should ensure that his/her staff comply with applicable laws and regulations for human subject protection and the conduct of clinical research.

The IEC/IRB is responsible for ensuring that the rights and welfare of study subjects are protected. The IEC/IRB should evaluate the adequacy of the site by considering the site's characteristics (e.g., number and qualifications of supporting staff, available facilities and equipment, and ability to cope up with emergency procedures).

It is the sponsors' responsibility to select appropriately qualified investigators to conduct the study and they should also assure they have sufficient staff (with appropriate qualifications) available, who are appropriately trained to conduct all study related activities and they understand how to capture and document the required observations and data.

The regulatory authorities may inspect the study sites to determine if the conduct of the study is in compliance with local laws/regulations. During these inspections they should find out who were assigned various responsibilities for conducting study-related activities (e.g., screening subjects to determine if they meet inclusion/exclusion criteria; obtaining informed consent; conducting physical examinations; collecting and analyzing study data; recording, transcribing or reporting data to the sponsor; administering the investigational product to subjects) and also determining whether these activities were appropriately assigned and within the scope of the staff member's professional license(s).

Principle 10: Records

All clinical trial information (raw data) must be recorded, handled and stored in a way that ensures that it is accurately reported, interpreted and verified. You must be absolutely certain that there is a clear data trail by documenting each and every decision at the time it is made.

Application

Principle 10 is applied through:

1. The understanding and application of basic elements of data quality and integrity
2. Adherence to the study protocol as well as applicable written procedures for collecting, recording, reporting, maintaining and analyzing clinical trial information
3. The preparation of essential documents (including source documents) at all stages throughout the conduct of the clinical trial.

Implementation

IEC(s)/IRB(s), investigators, sponsors and regulators have the responsibility for documenting their activities within GCP and maintaining records pertaining to duties related to the conduct or oversight of the clinical trial for the time required under national or local law and regulations. All parties are responsible for ensuring the accuracy, completeness, legibility and availability (as necessary) of such documents.

IEC(S/IRB(s) document their reviews of study protocols and informed consent/recruitment/advertising materials through minutes that capture the IEC(s)/IRB(s) deliberations and through copies of correspondence with the clinical investigator.

The investigators prepare and maintain case histories that record all observations and other data pertinent to the investigation on each individual who is administered the investigational drug or employed as a control in the investigation.

Sponsors ensure that study protocols address appropriate data handling and record-keeping requirements and design CRFs appropriately to facilitate the capture of all significant trial-related data and observations. Sponsors also secure the services of monitors who ensure compliance of the clinical investigators by verifying that the study was carried out according to the approved study protocol.

Regulators rely on clinical trial information to support regulatory decision-making and may inspect all of the parties involved in conducting or overseeing research. The regulators during inspection have direct access to clinical trial records for reviewing them. As part of an inspection, regulators compare records at the clinical investigator site and sponsor site with data and reports submitted to the regulatory authority to verify the information submitted. Regulators also prepare and maintain records of their inspections and findings.

Principle 11: Confidentiality / Privacy

The absolute confidentiality of study participants/subjects and their records must be protected so that the identity of subjects remains unknown to all except the PI (and her/his designates).

"The right of research subjects to safeguard their integrity must always be respected. Every precaution should be taken to respect the privacy of the subject, the confidentiality of the patients information and to minimize the impact of the study on the subject's physical and mental integrity and on the personality of the subject." (Declaration of Helsinki)

"The investigator must establish methods to safeguard the confidentiality of subjects' research data. Subjects should be told the limits, legal or other, of the investigators to safeguard confidentiality and the possible consequences of breaches of confidentiality." (CIOMS, International Ethical Guidelines, Guideline 18)

Application

Principle 11 is applied:

1. Through appropriate procedures to protect the privacy of the subject
2. By document and data control to protect the confidentiality of the subject's information.
3. Through the informed consent process which requires as an essential element that certain explanations be provided to the subject about the confidentiality of the subject's records and about access to those records by monitor(s), auditor(s), the IEC/IRB and the regulatory authority(-ies).

Implementation

IEC(s)/IRB(s) review/approve the informed consent procedures and document to ensure among other things, that there is adequate explanation regarding:

1. The risks related to release of the subject's private information
2. How the confidentiality of the subject's records will be maintained
3. Persons who may have access to the subject's records (e.g., monitor(s), auditor(s), the IEC/IRB and the regulatory authority(-ies).

Investigators should

1. Implement procedures to protect and restrict access to study records and private information (e.g., password protection for files, keeping study records in secure areas)
2. Follow national / local laws and regulations relating to privacy and confidentiality
3. Ensure that study staff are aware of and receive appropriate training related to their responsibility and procedures to be used for protecting subjects' private information and records
4. Ensure that study staff follow the procedures established for this study purpose
5. Ensure that the consent form and process inform study subjects about the procedures to be used to protect their private information and the circumstances under which their medical and study records may be viewed by regulators, sponsors, monitors and/or the IEC/IRB

Sponsors ensure that sites

1. Allow regulators, IEC(s) / IRB(s) and monitors direct access to records necessary to verify compliance with national / local laws and regulations pertaining to the conduct of clinical trials
2. Inform subjects and obtain their consent for such access

Regulatory authorities need to

1. Be alert to issues of subject confidentiality
2. Review sponsors, clinical investigators' and IECs' / IRBs' compliance with applicable national / local laws and regulations for handling private information and also informing subjects about these issues

Principle 12: Good Manufacturing Practice

Investigational products must be used only as per the protocol and must be manufactured, handled and stored in accordance with Good Manufacturing Practice (GMP).

"The sponsor should ensure that the investigational product(s)...is characterized as appropriate to the stage of development of the product(s), is manufactured in accordance with any applicable GMP and is coded and labeled in a manner that protects the blinding, if applicable..."(ICHE6, Section 5.13)

Application

Principle 12 is applied through:

1. Appropriately characterizing the investigational product (including any active comparator(s) and placebo if applicable)
2. Adhering to applicable Good Manufacturing Practice (GMP) standards in the manufacturing, handling and storage of the investigational product
3. Using the product according to the approved study protocol

Implementation

The responsibility for implementing this principle is shared by sponsors (or contract manufacturers / contract research organizations), investigators and regulators.

Sponsors implement this principle directly or indirectly through contract, by developing and characterizing the investigational product. They make the necessary notifications / submissions to the applicable regulatory authority (ies), identify GMP requirements, if any, that may apply to the manufacturing, handling and storage of the investigational product and ensure compliance with those requirements. Sponsors manufacture the investigational product directly or have it manufactured under contract at a manufacturing site in accordance with applicable GMP. They are responsible within GCP for the handling, storage, distribution and final disposition of the investigational product.

The sponsor also develops the study protocol and investigator's brochure, monitors protocol compliance and ensures that written procedures include instructions that the investigator / institution should follow for the handling and storage of investigational products for the trial and documentation thereof.

Investigators are responsible for familiarity with the investigator's brochure, and for conducting the research in compliance with the protocol, including any instructions for storing and handling investigational products. Investigators are responsible for explaining correct use (including handling and storage) of the investigational product to the study subjects. Investigators also ensure that any unused investigational products are returned to the sponsor after the trial is completed.

In accordance with national / local laws and regulations, regulators may establish GMP requirements for investigational products, review manufacturing data submitted in support of research permits or marketing applications and / or inspect manufacturing facilities. Since investigational products may be imported, regulators should be familiar with the manufacturing requirements in the country of origin and their conformance with international GMP standards.

Regulators may also inspect investigators for compliance with study protocol, including instructions for storing and handling investigational products.

Principle 13: Quality Systems

Systems/procedures that assure the quality of every aspect of the trial are to be implemented in a standard repeatable way such as...a clearly written, understood and authorized Standard Operating Procedures known in the industry as the "SOP's."

Application

Principle 13 is applied through development of procedures to control, assure and improve quality of data and records and the quality and effectiveness of processes and activities related to the conduct and oversight of clinical research.

Implementation

All the parties who conduct and oversee clinical trials (sponsors, clinical investigators, IEC(s) / IRB(s) and regulatory authorities) should adopt and implement quality systems for the processes and activities for which they are responsible.

Sponsors secure the services of the monitors to ensure compliance of the clinical investigators and verify that the study was carried out according to the approved study protocol. Sponsors also audit the monitors' performance and other quality control activities and systems to ensure each system's performance.

Monitors review study records at the sites, report their findings to the sponsor and prepare written reports that document each site visit or trial-related communication.

Investigators through supervision ensure that study staff follow established procedures for the conduct of the study (e.g., obtaining IEC / IRB approval of the study, obtaining informed consent from subjects, establishing and maintaining subjects' case histories, transcribing data from subjects' medical files to the CRFs, reporting adverse events and other unanticipated problems, etc).

IEC(s) / IRB(s) develop and adopt SOPs for reviewing studies and informing the clinical investigator of any required modifications to the study protocol and for assuring such modifications are in place before the study proceeds. In accordance with national / local laws and regulations, IEC(s) / IRB(s) may develop SOPs to allow IEC / IRB members or a third party to observe the consent process to verify that subjects are being provided the opportunity to ask questions about the study and that subjects receive a copy of the informed consent document. IEC(s) / IRB(s) implement systems to assure that continuing review of the study takes place at intervals appropriate to the degree of risk and that investigators are notified so that they may provide the necessary documentation to IEC /IRB in advance of the deadline.

In accordance with the applicable laws / regulations, regulators may inspect all parties that conduct or oversee research and verify the information submitted to the regulatory authority. Regulators may ask for sponsors' monitoring plans as a condition of allowing a study to proceed. Regulatory authorities also optimally develop SOPs and quality systems for internal regulatory activities, including policies and procedures for reviewing product applications and for the conduct of GCP inspections.

Suggested Reading

1. Alain Spriet. GCP of clinical drug trials. Karger Publication.
2. Gerhand Fortwegel. Guide for clinical research staff. Karger publication.
3. GCP for clinical research in India. CDSCO publication.

CHAPTER 14

Role and Responsibilities of Sponsor

U. Shobha Jagadish Chandra

14.1 Introduction

Sponsor may be an individual or a company, institution or organization which takes the responsibility for the initiation, management and/or financing of a clinical trial.

Quality Assurance (QA): All those planned and systematic actions that are established to ensure that the trial is performed and the data are generated, documented (recorded), and reported in compliance with good clinical practice (GCP) and the applicable regulatory requirements.

Quality control (QC): The operational techniques and activities undertaken within the quality assurance system to verify that the requirements for quality of the trial related activities have been fulfilled.

The sponsor is responsible for implementing and maintaining quality assurance and quality control systems with written SOP's to ensure that trials are conducted and data are generated, documented (recorded) and reported in compliance with the protocol, GCP and applicable regulatory requirements.

14.2 Contract Research Organization (CRO)

A sponsor may transfer any or all of the sponsor's trial related duties and functions to a CRO but the ultimate responsibility for the quality and integrity of the trial data always resides with the sponsor. The CRO is responsible for implementation of quality assurance and quality control. CRO provides a growing proportion of skilled staff and facilities involved in clinical research around the world. The sponsor CRO realignment has increased the speed and efficiency of drug development efforts while maintaining clinical trial data quality and high level regulatory compliance.

Any trial related duty and function that is transferred to and assumed by a CRO should be specified in writing and any of those duties and functions not specifically transferred to the CRO are retained by the sponsor. All references to a sponsor in this guidelines also apply to a CRO to the extent that a CRO assumes the trial related duties and functions of a sponsor.

The sponsor should designate appropriately qualified medical personnel who will be readily available to advise on trial related medical questions or problems. If necessary, external consultants may be appointed for this purpose. The sponsor should utilize qualified individuals (eg. Bio-statistician, clinical pharmacologist and physicians) as appropriate throughout all stages of the trial process, from designing the protocol and CRF's and planning the analyses to analyzing and preparing interim and final clinical trial reports.

The sponsor may consider establishing an independent data monitoring committee (IDMC) to assess the progress of a clinical trial, including the safety data and the clinical efficacy end points at intervals and to recommend to the sponsor whether to continue, modify, or stop a trial. The IDMC should have written operating procedures and maintain written records of all its meetings.

If electronic handling of the trial data is being considered the sponsor has to ensure

1. The electronic data processing system conforms to the sponsor's established requirements for completeness, accuracy, reliability, and consistent intended performance.
2. Maintain SOP's for using these systems.
3. Ensure that the systems are designed to permit data changes in such a way that the data changes are documented and that there is no deletion of entered data (maintain an audit trial, data trial, edit trial)
4. Maintain a security system that prevents unauthorized access to the data.
5. Maintain a list of the individuals who are authorized to make data changes
6. Maintain adequate back up of the data.

7. Safeguard the blinding, (maintain the blinding during data entry and processing. If data are transformed during processing it should always be possible to compare the original data and observation with the processed data. The sponsor should use an unambiguous subject identification code that allows identification of all the data reported for each subject.

The sponsor has to retain all the essential documents pertaining to the trial in conformation with the applicable regulatory requirements of the country where the product is being approved and/or where he intends to apply for approval. These documents have to be retained for a longer period if required by the applicable regulatory requirements. Trial documents have to be retained until at least 2 years after the last approval of a marketing application. The need for retaining or destroying the records should be notified to the investigator or institutions in writing. In case the sponsor discontinues the clinical development of an investigational product it is his responsibility to maintain all the essential documents for at least two years after formal discontinuation or in conformance with the applicable regulatory requirements. If the clinical development of the investigational product is discontinued then he has to notify all the trial investigators/institutions and all the regulatory authorities.

As required by the applicable regulatory requirements, any transfer of ownership of the data should be reported to the appropriate authorities. The most important responsibility of the sponsor is selecting the investigator/institution. The investigator should be selected based on the qualification, training and experience, and have adequate resources to properly conduct the trial. If there is a need to conduct a multi-center trial the selection of organization and the co-ordinating investigator is also the sponsor's responsibility. Once the investigator is selected, sponsor has to provide the protocol and up-to-date investigator's brochure (which contains safety and efficacy data from non-clinical studies and or clinical trials to support human exposure by the route, at the dose, duration and the trial population studied) with sufficient time to review the information provided.

After the investigator reviews the data provided, the sponsor should obtain the investigator's/institute's agreement. The agreement should contain all the points pertaining to the trial like the trial to be conducted as per GCP guidelines, agreement with the regulatory requirement, the protocol agreed upon and favourable opinion by the IRB/IEC. All procedures for recording and reporting will be in compliance with the protocol. Permit monitoring, auditing and inspection during the study period. Retain all the trial related essential documents until the sponsor informs that they are no longer required. The financial aspects of the trial should be documented as per the agreement. The sponsor and the investigator/institute should sign the protocol, or an alternative document to confirm this agreement. Prior to initiating a trial the sponsor should define, establish and allocate all trial related duties and functions.

The sponsor should provide insurance or indemnify (legal and financial coverage) to the investigator/institution against claims arising from the trial, except for claims that arise from malpractice and or negligence. The sponsor's policies and procedures should address the costs of treatment of trial subjects in the event of trial related injuries in accordance with the applicable regulatory requirements. The method and manner of compensation to the trial subjects should comply with applicable regulatory requirements.

The appropriate authorities have to review and give acceptance or permission to begin the trial. Any notification/submission should be dated and contain sufficient information to identify the protocol.

The sponsor should obtain the following information from the investigator / institution

(a) The name and address of the investigator/institution's IRB/IEC.

(b) A statement from IRB/IEC that it is organized and operates as per GCP applicable laws and regulations.

(c) Documented IRB/IEC approval/favorable opinion. If IRB/IEC conditions its approval such as modifications of protocol, written informed consent form and any other written information to be provided to the subject and or other procedures, the sponsor should obtain from the investigator/ institution a copy of the modification made and the date of approval/ favorable opinion given by the IRB/IEC.

14.3 The Sponsor and the investigational Products

(a) The responsibility of the sponsor lies in ensuring that the investigational product and placebo are characterized as appropriate to the stage of development of the product and is manufactured in accordance with any applicable GMP and is coded and labeled in a manner that protects the blinding??. Labeling should comply with the applicable regulatory requirements.

(b) Sponsor should determine that the acceptable storage temperatures, storage conditions like protection from light, storage times, reconstitution fluids and procedures and devices for product infusion are available at the study site. This information has to be passed on to the monitors, pharmacist and storage managers.

(c) Packing of the investigational product to prevent contamination and unacceptable deterioration during transport and storage is another aspect which the sponsor has to look into.

(d) In blinded trials the coding system for the investigational product should include a mechanism that permits rapid identification of the products in case of a medical emergency but does not permit undetectable breaks of the blinding.

(e) If significant formulation changes are made in the investigational or comparator products during the course of clinical development the results of any additional studies of the formulated products.(eg. stability, dissolution rate, bioavailability) needed to assess whether these changes would significantly alter the pharmacokinetics profile of the product and should be available prior to the use of the new formulation on clinical trials.

(f) The sponsor should ensure that written procedures should address adequate and safe receipt, handling, storage, dispensing, retrieval of unused product from subjects and return of unused investigational products to the sponsor (or alternative disposition if authorized by the sponsor and in compliance with the applicable requirements.

(g) Timely delivery of the investigational product to the investigator is another responsibility of the sponsor.

(h) Maintain records that document shipment, receipt, disposition, return and destruction of the investigational products.

(i) Maintain a system for the disposition of unused investigational products and for documentation of this disposition.

14.4 Sponsor and Record Access

The sponsor is responsible for ensuring that a mention is made in the protocol or in the agreement that the investigator/institute provide direct access to source data/documents for trial relating monitoring, audits, IRB/IEC review and regulatory inspection.

14.5 Safety Information

The sponsor is responsible for on going safety evaluation of the investigational product. The sponsor has to notify all concerned investigators/institutions and the regulatory authorities of findings that could affect adversely the safety of subjects, impact the conduct of the trial or alter the IRB/IEC's approval/favourable opinion to continue the trial. Any serious and unexpected adverse drug reactions have to be reported to the investigator/institution, IRB/IEC and to the regulatory authorities. All the safety updated and periodic reports have to be submitted to the regulatory authorities.

Suggested Reading

1. Barry Bleidt. Clinical research in pharmaceutical development. Marcel Dekkar Publication.
2. International conference on harmonisation of technical requirements for registration of pharmaceuticals for human use. ICH guidelines-Clinical safety data management. Oct 1994.

CHAPTER 15

Role and Responsibilities of the Principal Investigator

P. Usha Rani and M.U.R. Naidu

15.1 Introduction

A principle investigator is an individual who assumes full responsibility for a research project, including the supervision of any research assistants and students. He or she, is responsible for the conduct and reporting of scientific design. Additionally, the Principal Investigator is responsible for the direction, oversight, conduct, financial management, successful completion, and reporting of the project, for managing any collaborative relationships and be qualified by training and experience. International guidelines specify that, the medical care and decisions made for trial subjects should be the responsibility of a qualified physician or dentist where applicable. While technically a non-physician can be listed as PI, a physician is typically included as a sub-investigator to perform medically or clinically related study functions.

15.2 Competencies and Responsibilities of the Principal Investigator

Prior to commencement of the trial, the PI must

- submit written undertaking to conduct the trial as per specified protocol to sponsor, regulatory authorities, IEC

- ensure that all documents required for approvals from the local institutional ethics committee, are received from the sponsor and submitted to ethics committee
- all regulatory authorities approvals if applicable, are obtained
- if investigator is conducting a multinational study, Form 1572 must be signed before an investigator may begin participation in a clinical study conducted under the IND regulations. The investigator should sign this form only AFTER having been given enough information to be informed about the study and to understand the regulatory commitments made via the form. This typically means the investigator must have received, read, and understood the investigator brochure and protocol, and be familiar with the regulations before signing a 1572.
- ensure the duly signed trial agreement and essential documents are obtained
- have good knowledge of the protocol , related documents (indemnity and insurance)
- review carefully and sign study documents including clinical trial agreement
- have read, understood and agreed to work according to the protocol, the Declaration of Helsinki, ICH Guideline for Good Clinical Practice and other National relevant guidelines and legislation;
- undertake to use the investigational and comparator products only for the purposes of the study as described in the protocol;
- document clearly the sequence of events to be followed in the conduct of the clinical trial, including timeframes, roles and responsibilities delegated to study personnel;
- ensure the availability of all necessary facilities, equipment, and finance to conduct the trial;
- develop proper mechanisms to ethically obtain informed consent of participants;
- assure that labs and work places are identified with appropriate warning signs; and that personnel are properly trained and are informed regarding biohazards;
- inform study personnel of the reasons and provisions for any precautionary medical practices advised or requested, such as serum collection etc;
- prepare a written contingency plan for handling emergency conditions such as emergency medical resuscitative procedure, personnel exposure to contaminanta, loss of containment, power failure, etc.
- accept the involvement of monitors to review and verify the quality control procedures and conduct data verification;

- accept the possibility of audit and/or inspection by an independent auditor appointed by the sponsor, regulatory authority or ethics committee;
- should prepare and obtain adequate budget from sponsor to conduct the study (Ref Budget checklist)
- obtain the right to publish; it is unethical for the sponsor to reserve to right to publish;
- generate the information package for the participant, and where applicable with the sponsor;
- ensure proper safety reporting procedures.

During the trial PI must

- Ensure informed consent is taken from all study participants before any study related procedures
- Reconsent to be obtained as and when required for ongoing studies on IEC approved consent forms
- Ensure proper maintenance of trial master file and other essential study documents securely
- Adequate and proper source documents are maintained
- Ensure confidentiality of study and study participants is maintained
- Staff and patients comply to study protocol
- Ensure time lines are being met
- Ensure regular updates are intimated to IEC on trial progress
- In case of SAE, ensure timely updates are given to sponsor and ethics committee
- Ensure timely and regular supply of study material is obtained from sponsor
- Take responsibility for accountability of the investigational products
- Ensure all study related documents and products are stored in a secure storage area
- Ensure patients come for regular and timely follow ups as per protocol
- Regular evaluation of efficacy parameters as per protocol
- Monitor patients safety regularly
- Permit monitoring of trial and related documents by study monitors, auditors and regulatory authorities

Before the project ends (best practice = 3 months prior to the end date), the PI should

- Assure that subrecipients have submitted or will submit all required deliverables, reports and invoices.
- Plan for updates and changes to payroll, transitioning either to a renewal period or to award expiration

- Request Preaward Spending for a renewal or continuation period when appropriate
- Request a no-cost extension when allowable and appropriate

When a project expires or is terminated, the PI should

- Meet sponsor's deadline(s) for final reports
- Submit any required final project report
- Submit any required final invention report
- Review the final financial report
- Destroy or return investigational products and study related material as per agreement with sponsor
- Report Close out to IRB
- Retain the project records in archives for as long as they are required as per local regulatory requirement.

15.3 Qualifications and Availability

The investigators should be qualified by education, training, and experience to assume responsibility for the proper conduct of the trial, should meet all the qualifications specified by the applicable regulatory requirements, and should provide evidence of such qualifications through an up-to-date curriculum vitae and/or other relevant documentation requested by the sponsor, the ethics committee, and the regulatory authorities. The investigator must have a suitably qualified and experienced team to whom significant trial-related duties may be delegated (Refer to Sample delegation log). Further, the investigator must have sufficient time to spend with the trial participants, staff and sponsor's monitors and auditors as applicable.

15.4 Adequate Resources

The investigator should have the potential for recruiting the required number of suitable participants within the agreed recruitment period and also have sufficient time to properly conduct and complete the trial within the agreed trial period. Further, there should be an adequate number of qualified staff and adequate facilities for the foreseen duration of the trial to conduct the trial properly and safely. The investigator should ensure that all persons assisting with the trial are adequately informed about the protocol, the investigational products, and their trial-related duties and functions.

15.5 Informed Consent

The PI is responsible for ensuring that an adequate information package, in an acceptable format, is available for use in the process of seeking informed consent from participants to participate in the study. The informed consent form approved by the local institutional ethics committee should be used. In all instances both written and verbal informed consent should be obtained. Verbal consent, where the participant is illiterate, should be obtained in the presence of and countersigned by a literate witness. Further, the PI, co-investigator, or designated person as defined in the protocol, should then seek the participant's informed consent to participate in the study in accordance with the principles outlined in the Declaration of Helsinki, and other relevant National Regulatory guidelines. After consent to participate in the study has been obtained, the original signed informed consent form should be kept with the trial records and a copy of signed informed consent form should be provided to the participant. The details of the consent procedure should be documented in source notes including the questions asked by the participants and the answers given by the investigator.

If the study participant is a referred subject, the principal investigator, should inform the medical practitioner about inclusion of the respective subject in a trial.

15.6 Investigational Products

Accountability of the investigational products at the trial sites is the responsibility of the PI. The investigator and/or a pharmacist or other appropriate individual, who is designated by the PI, should maintain records of the product's delivery to the trial site, the inventory at the site, the use by each subject, and the return to the sponsor or alternative disposition of unused products. These records should include dates, quantities, batch/serial numbers, expiration dates (if applicable), and the unique code numbers assigned to the investigational products and trial participants. The envelops containing the blinding codes should be maintained by the PI or the designee in a secure cupboard. Investigators should maintain records that document adequately that the participants were provided the doses specified by the protocol and reconcile all investigational products received from the sponsor. The investigator must ensure that the investigational products are used as per the protocol and by the end of the study, all the unused medication are returned to the sponsor or disposed off as per the investigating site SOPs.

15.7 Compliance with Protocol

It is the investigator's responsibility to conduct the study as per the protocol approved by the local ethics committee and regulatory body (if applicable). Any changes made to the protocol needs prior review by sponsor and approval by ethics committee. Incase of any deviation, the investigator, or person designated by the investigator, should document and explain such changes to the approved protocol.

In an emergency situation, if the investigator has implemented a change to the approved protocol, the implemented deviation or change, the reasons for it, and, if appropriate, the proposed protocol amendments should be documented as soon as possible and submitted:

(a) to the ethics committee for review and approval,

(b) to the sponsor for agreement and,

(c) to the regulatory authority.

15.8 Monitoring and Auditing

The relationship between the principal investigator, the monitor and the sponsor must be clearly defined and stated in writing in the study agreement or related documents which must be stored in a secure place. The investigator must permit periodic inspections and monitoring by the sponsor and further, accept the possibility of an audit and/or inspection by the sponsor, regulatory authority or ethics committee. The investigator must be available for agreed visits by the monitor during the study and also co-operate in the data editing, quality control and audit.

15.9 Source Data and Other Documentation

The principal investigator is responsible for the collection, quality, recording, maintenance and retrieval of source data arising from the clinical study which is transcribed from source documents to a structured case record form. The source document must be signed and dated by the investigator or designated person, on a visit by visit basis and then stored securely, but made available to the sponsor/ monitor on request and for auditors and/or regulatory authorities for audits and inspections.

However, corrections to CRFs can only be made by the principal investigator, co-investigator or designated person where a single line should be drawn through the data in such a way that the original entry is not obscured, and the correct data inserted nearby. All corrections should be initialed and dated by the corrector.

For a web-based data entry system, direct data entry onto a computer should only be entered by the investigator and or a designated person. The computer system should be virus and tamper proof .with restricted access. An audit trail must be available to record the data changes which are documented so that there is no deletion of entered data.

15.10 Safety Issues and Reporting of Serious Adverse Events

It is the investigator's responsibility to ensure that adequate provisions are made for dealing with any serious and /or unexpected adverse events that may occur to the study participants as per the SOPs of the sponsor or the investigating site. It is advisable to develop treatment guidelines for the clinical management of some adverse events. These operating procedures should be included in the protocol and its related documents.

The principal investigator must inform the sponsor of any serious or unexpected adverse events occurring during the study within the time specified in the protocol using the defined reporting forms, A serious adverse event initial report form and any relevant follow-up information should be sent to the sponsor , to the appropriate ethics committee, and regulatory authority.

15.11 Premature Termination - Breaking the Treatment Code

In a blinded, randomized study, the investigator should ensure that the binding is maintained and code is broken only in accordance to the procedure as stated in the protocol and known by the staff involved in the clinical management of the participants. The principal investigator and/or a designated person should keep the treatment code list, code-break envelopes or code-break cards accessible 24 hours a day at the study site in the event of an emergency. Further, the investigator should promptly document and explain to the sponsor any premature withdrawal of the investigational products which in turn should be reported to the regulatory authority by investigator via the sponsor where there is a sponsor, in writing and within 24 hours of the incident.

15.12 Progress Reports, Trial Outcome and Final Study Reports

The investigator is obliged to submit periodic progress reports to the ethics committee and as required to the sponsor and/or, the regulatory authority if applicable. These reports should contain information on how the study is progressing, the number of participants included in relation to the number expected, the number of dropouts and withdrawals and if the planned time schedule is still appropriate. After final data management, the results are analysed and a final report on completion of the study should also be submitted in the prescribed format to the regulatory authority. A copy of the same is archived by the investigator securely.

Suggested Reading

1. Spilker B. Guide to clinical trials. New York 1991. Raven Press.
2. FDA Form 1572 - Guidance on the FDA's Statement of Investigator Form.
3. Charles Y C Park. Techniques of patient oriented research. Raven Press.
4. Marcus Plather. Hand book of clinical trials. ReMedica Publication Sample Investigator Delegation of Task Authorization

Simple Investigator Delegation of Task Authorization

Protocol:____

Sponsor: ____

Protocol Number:____

Site Name: ____________________ Site Number:____________________

PrincipalInvestigator: ___________________________________

Protocol Procedures for Reference:

1. Informed Consent Process
2. Physical Exam(s)
3. Perform EKG
4. Read EKG
5. Adverse Event Reports
6. Adverse Event Assessment
7. Dispense study medication medication/Compliance check
8. Lifestyle and diet counseling
9. Medical History
10. Collect Routine Safety Labs
11. Evaluate lab results
12. Process and ship lab samples
13. Quality of Life Questionnaire
14. Concomitant Medication Log
15. Collect study
16. Routine vitals

I hereby authorize the following personnel at __________ to perform the following tasks:

List name, signature, initials and task

Name Print	Credential	Title or Role	Signature	Initials	Procedure(s)List numbers

Principal Investigator Signature: ______________________________

Date: __________

Check list for Clinical Trials Budgets

Expenses	Considered	Amount Estimate in Rs. /Comments
Review of Protocol	☐	
Review/prepare/modify informed consent	☐	
Nursing • Transcribing source documents • Diary review • Immunization • Other	☐	
Physician • Exam • Other	☐	
Research Assistant • Scheduling visits • Clinic preparation • Clinic receptionist • Telephone reminders • Other		
Pathology fees	☐	
Patient travel / costs / parking	☐	
Pharmacy fees	☐	
Physical exam	☐	
Communication with Sponsor	☐	
Prepare – UBC Ethic Review Board & C&W Research Review Submissions	☐	
Printing • Consent forms • Material explaining study • Immunization records • Letterhead • Label mail outs	☐	
Transcribing source documents	☐	
Recruitment • Advertising • Letters	☐	
Office supplies – copying, labels, etc	☐	
Laboratory fees	☐	
Completion of case report forms (CCRF)	☐	
General equipment / supplies	☐	

Expenses	Considered	Amount Estimate in Rs. /Comments
Advertising	☐	
Clinic space rental	☐	
Data storage	☐	
Electrical set up	☐	
Equipment maintenance	☐	
Quality assurance review	☐	
Patient Record Retrieval	☐	
External Legal Fees	☐	Budget will be provided by external legal counsel
Staff training	☐	
Shipping	☐	
Postage	☐	
Internal review board fees (IRB)	☐	
Overhead fixed 25%	☐	
Other	☐	

Additional Notes:

If there are any budget queries, please identify the contact Principal Investigator.

Name: ______________________________

Phone: ______________________________

Ethical Considerations in Clinical Research - Special Reference to Research in Developing Countries

P. Usha Rani and M.U.R. Naidu

16.1 Introduction

"Research should not be carried out in developing countries just because it is cheaper and the laws are more lax. The same ethical rules should apply where ever research is being conducted." Dr.Anders Milton, Chairman, World Medical Association Assembly, Edinburgh, Scotland. (Oct 2000)

Protecting the rights and welfare of those who volunteer to participate in research is a fundamental tenet of ethical research. The studies in question should address important health concerns faced by the country. Sometimes simple alternatives exist, such as conducting a literature search etc. To conduct a clinical trial that is not truly necessary is contrary to commonly accepted standards, places patients at unnecessary risk, and may be considered unethical.

180

The research community and the public have been engaged in a reexamination of the ethics and responsible conduct of research involving human subjects for many years. Attention has recently focused on research carried out in countries where the requirements for review, approval, and oversight are much less stringent than they are in the United States. Some studies have exploited the unfortunate conditions in which some people live as a result of economic and cultural factors beyond their control — factors that make them vulnerable and convenient for study.

This chapter aims to address the principal ethical requirements surrounding the conduct of clinical research as follows:

1. Individuals involved with ethical issues
2. Ethical considerations in protocol development including the choice of research designs, especially where placebo is proposed
3. Issues arising in informed consent procedure
4. Considerations during conduct of the trial
5. Benefits provided to research participants after the trial has ended and
6. Finally the need for developed countries to assist developing countries in building the capacity to become fuller partners in international research

16.2 Individuals Involved With Ethical Issues

The four pillars which lay the foundational concept of ethics in clinical trials surrounding the trial subject are:

(a) The principal investigator and clinical trial staff
(b) The institutional review board / ethics committee
(c) The sponsor and staff of sponsoring institution
(d) Regulatory agency

Key Players

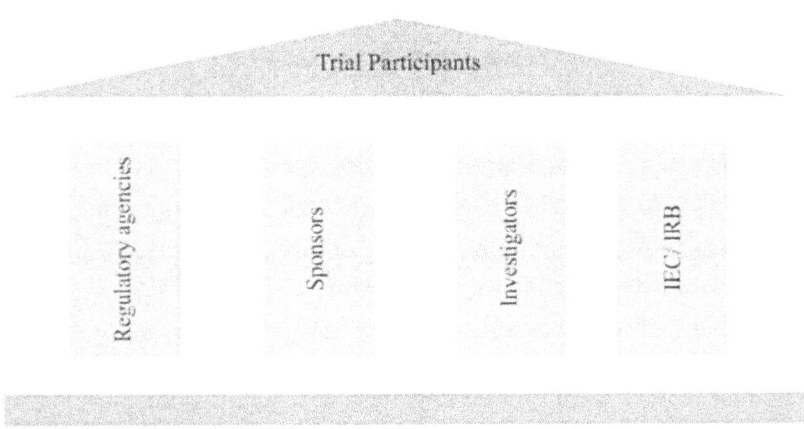

(a) The research study. Being physicians, the clinical investigators are responsible for maintaining highest ethical standards of practice. The specific ethical responsibilities assigned to them include: (1) To ensure whether the protocol adheres to the normal acceptable procedures; (2) Protocol is approved by the relevant ethics committee; (3) Ensure patient gives valid informed consent before any trial related procedure and finally (4) Protocol is followed according to appropriate and proper medical practice. The investigating staff are assigned specific study related responsibilities by the principal investigator. The ethical issues here include how the staff carryout their duties as protocol violations caused by them range from minor incidents to serious negligence. Other site personnel including those handling biological samples must follow commonly accepted medical standards in discharging their responsibilities.

(b) The IRB/ethics committee acts as the major ethical overseer of the clinical trial. It is now widely accepted that research involving human subjects should be conducted only after an appropriate ethics review has occurred. Ethics review committees must assess the justification provided, including the risks to participants, and overall ethical acceptability of the research design. Further, the IRB reviews the periodic reports furnished by the investigator on the progress of the trial and on any serious or unforeseen adverse events.

(c) It is important for a sponsor to choose an investigator who will maintain high standard of research, otherwise the entire ethical conduct of the study may be severely compromised. Sponsor's staff monitoring the trial should also be cognizant of the standards followed by the investigating team and patients during the conduct of the study.

(d) The regulatory agency reviews whether all the protocols submitted are ethically sound. They also ensure that appropriate standards are maintained; thus safeguarding the review process. Thus the various regulations promulgated by the agency govern the various steps in protocol approval and influence the numerous ethical issues throughout the design and approval process.

16.3 Ethical Issues Related to Protocol Development and Study Design

Clinical research conducted in developing countries should be limited to those studies that are responsive to the health needs of the country.

During **protocol development**, selection criteria of study participants especially trials involving vulnerable groups including children, pregnant women etc may raise ethical questions on how these criteria are applied. The lower age limit set for patient enrolment in adult trials is an ethical issue to be addressed.

Further, in studying most investigational products, all women of child bearing age are excluded and if a female participant becomes pregnant, she must be immediately discontinued from the trial and followed up during the entire course of pregnancy to evaluate the new born infant.

There may be ethical reasons to exclude patients from clinical trials who were previously treated with certain medications. Sometimes a washout or medicine free period is instituted whenever patients are switched from one treatment to another. An ethical dilemma may be created as many patients with numerous diseases cannot tolerate drug free periods. Perhaps this can be overcome if the protocol allows investigator to eliminate consideration of the first several weeks on the new medicine because of the possible carryover effect. If trial participants engage in unethical practice of enrolling in multiple medicine trials at the same time, residual effect would confound the data obtained. Further, modification of dosage scheduled may be required for safety reasons; hence it is ethically desirable to incorporate a systematic plan in the protocol itself to this effect.

There are usually limits on the amount of resources available to conduct any clinical research. At times, it may be medically desirable but financially impossible to utilize specific equipment. Thus, if some or all important parameters cannot be measured then it may not be ethically justified to conduct the study. Further, if any toxicity is anticipated, then specific attempts should be made to evaluate patients at risk. Finally, handling of biological samples especially where genetic studies are involved, must be addressed specifically while designing protocols.

Choosing an appropriate **research design** for a clinical study depends on various contextual considerations, so, what may be an ethically acceptable design in one situation may not be so in another. Researchers should thus provide the ethics committee a thorough justification of the research design to be used, including the procedures to be used to minimize the risk to participants.

To a large extent, the appropriateness of a clinical trial design and of the control group used is based on two considerations. First, the research question should be posed in a manner that is relevant and meaningful to the host country, such as a comparison of an available treatment with the current standard of care in the host country (which might be no intervention in some cases). This approach may be necessary in order to avoid a situation in which the findings of the research are of interest to, or relevant to, only the scientists and health policy officials in the developed country sponsoring the study and would hold no promise for improving health in the host country. Second, the study should be designed to achieve maximal efficiency in producing meaningful data that could lead to improvements in health in the host country without leaving the study participants worse off at the end of the trial than they were when they entered it.

From the perspective of the protection of human participants in research, one of the most critical issues in clinical trial design concerns the use and treatment of control groups, which often are an essential component in methodologies used to guard against bias. These type of studies are called active-control (or positive control) studies, which are often extremely useful in cases in which it would not be ethical to give participants a placebo because doing so would pose undue risk to their health or well being. Thus researchers and sponsors should design clinical trials that provide members of control group with an established effective treatment. Any study that would not provide the group with an established effective treatment should include a justification for using an alternative design. The ethics committee must assess the justification provided as well as the overall acceptability of the study design. Another question to be answered here is, Are the patient's chances of receiving an active medication acceptable? That is, all patients should have at least an equal chance of receiving an active medicine rather than a placebo. It may be ethically unacceptable to enroll and treat patients in a trial if the chances of receiving active medicine are relatively low.

The ethical acceptability of including a **placebo** treatment group in a clinical trial is often considered. Many trials, however, are preceded by a "placebo run-in", in which all patients are given placebo. This is a commonly method employed to weed out non-compliers before randomization, to eliminate placebo responders, ensure patients are stable, washing out previous treatment etc. This poses further ethical and logical problems. Thus, an experimental intervention should normally be compared with an established, effective treatment (defined as a treatment that has widespread acceptance by the medical profession throughout the world and that is as effective as any alternative treatment for the disease or condition). Therefore, the presumption is that a placebo control, or any other control that is less effective than an established, effective treatment, is not ethically acceptable.

For clinical research to be ethical, it must fulfill several universal requirements. Among other requirements, it must be scientifically valid and must minimize the risks to which the research participants are exposed. There is a middle ground when placebo is advocated in a clinical trial. First, both sides agree that certain placebo-controlled trials are clearly unethical. If effective, life-saving, or at least life-prolonging treatment is available, and if patients assigned to receive placebo would be substantially more likely to suffer serious harm than those assigned to receive the investigational drug, a placebo-controlled trial should be prohibited. Second, advocates of active controls should agree that for ailments that are not serious, if there is only a minimal chance that patients randomly assigned to receive placebo will suffer harm or even severe discomfort, the use of placebo controls is ethical. A placebo-controlled trial of a new treatment for allergic rhinitis would be ethical because the moderate discomfort associated with allergic rhinitis typically does not impair health or cause severe discomfort. Indeed, the risks associated with such trials are no greater than those deemed acceptable in natural-history and epidemiologic studies in which blood samples are obtained solely for research purposes and in pharmacokinetic studies in which medications are administered to healthy volunteers and blood samples obtained from them even though there is no prospect of a benefit to the study participants.

Thus, Placebo-controlled trials are caught in a battle between two orthodoxies. One is that placebo should be used as a control unless there is an increased risk of death or irreversible morbidity associated with its use. The other view is that if an effective therapy exists, the use of a placebo should be prohibited. These two positions are both absolute and indefensible. Hence a middle ground in which placebo-controlled trials are permitted but only when the methodologic reasons for their use are compelling, a strict ethical evaluation has made it clear that patients who receive placebo will not be subject to serious harm, and provisions have been made to minimize the risks associated with the receipt of placebo. This framework provides a basis for deliberation in difficult cases, with the recognition that reasonable people might make divergent judgments in a particular case.

It is also the goal of any clinical trial to have a high **power**. If a trial is designed with low power, it may be unethical to enroll and treat patients in such a trial. Thus sufficient attention has to be given to **sample size** while designing a good clinical trial.

16.4 Issues on Informed Consent

The requirement to obtain voluntary informed consent from human participants before they are enrolled in research is a fundamental tenet of research ethics. Nuremberg Code in 1947 was the first to proclaim this requirement, which appeared in all subsequent published codes, regulations and guidelines pertaining to research ethics. From an ethical perspective, the informed consent is one of the most significant issues relating to a clinical trial. By signing an informed consent, a patient accepts the responsibility of complying with the terms of the trial. Researchers should strive to ensure that individuals agree to participate in research without coercion or undue inducement. Special ethical considerations are raised in trials performed on infants, children, pregnant women, prisoners, mentally retarded, psychotic or other individuals who are unable to protect their own rights fully. A major ethical problem encountered with these groups relates to the difficulty of obtaining a freely given informed consent.

The Ethics committee should review and approve all the informed consent documents including translations in local languages. Researchers should develop procedures to that potential participants do, in fact, understand the information provided in the consent process and should describe those procedures in their research proposals.

It may be important to tailor the particular procedure for obtaining voluntary informed consent in developing countries according to local custom and culture of the developing countries where a variety of religious groups, social class and languages exist. Cultural dependence on the decision of the head of the family or village/community leader further complicates the matter.

In the developed countries, the regulatory procedures focus more on the informed consent document rather than the process of informed consent and require written consent. In developing countries where the subjects may be illiterate, it may be difficult to implement such procedures. In such situations, an ethically sound alternative to written consent is oral consent that has been witnessed and verified.

16.5 Ethical Issues During Conduct of a Clinical Research

The various aspects of a clinical trial from recruitment of subjects to termination of a subject have ethical implications. An important safeguard is needed to avoid the exploitation of potentially vulnerable populations in developing countries. It is unethical to offer a relatively large amount of money or other non medical inducements to a patient with the primary motivation to enroll in a trial. On the contrary, it is also unacceptable to charge patients for some clinical trials with medical devices.

Further, collecting either too little or too much data in a clinical trial also raises ethical questions. For too little data may be inappropriate for any statistical analysis and too much data collected may actually interfere with the trial conduct.

All patients enrolled in a clinical trial should be given a card indicating they are in a trial and the name, contact address and phone number of the investigator and/or coordinator. An issue arises when a patient is in a double blind trial and a medical emergency develops. Likewise, after a patient is discontinued from a clinical trial because of a medical problem, it is the responsibility of the physician in charge of the trial to continue care till the patient's condition has returned to normal state or transferred to the care of another physician. It is likewise the responsibility of the investigator to obtain contact information of patients lost to follow up. There could be an ethical problem if the patient had to be contacted during or after the trial about an issue that could affect their health.

Ethical questions concern not only which population of patients is permissible to study and which trial designs are ethical to conduct but also when it is no longer ethically acceptable to continue a trial to the scheduled completion. The means used in terminating a trial may raise additional ethical concerns. A trial may be discontinued early for negative reasons like development serious adverse drug reactions, failure of investigating staff to maintain ethical standards; decision by sponsor in drug development etc or for positive reasons like major beneficial effect of treatment is observed that raises ethical concerns for patients who are not receiving this treatment.

Another issue faced in developing countries is lack of proper documentation during clinical trials and proper storage of trial related documents. The research team is not stable and trial documents are not user friendly. Thus leading to difficulties in obtaining uniformity of data in multicentric trials.

Biological samples like blood, cells, biopsy specimens etc are collected by researchers during a research project. Today, new technologies and advances in biology provide more effective tools for using such resources to improve medicine's diagnostic and therapeutic potential. The very power of these new technologies raise a number of important ethical issues. First question which arises is, whether it is appropriate to use stored biological materials in ways that originally were not contemplated. Secondly, the extent to which a research sample can be linked with the identity of its source is an important determining factor to assess the risks and potential benefits to the trial participants.

Finally, researchers have a responsibility to make sure that public is accurately informed about results without raising false hopes or expectations. Investigators should take care to maintain confidentiality while publishing data.

16.6 Access to Post Trial Benefits

An important ethical consideration in clinical trials of investigational medicines is: "Will the patients who have benefited from treatment be allowed to continue receiving the medicine after the trial is completed"? Ethics review committees should require that researchers include in the informed consent process and consent documents information about what benefits, if any, will be available to research participants when their participation in the study in question has ended. Further, in early discontinuation trials, investigator must inform patients in some manner about the results, arrange for continuation of their medical care, collect the final data to be incorporated into the data collection forms.

Ethical issues arise at the conclusion of all clinical trials. The Declaration of Helsinki states (Principle 30), "At the conclusion of the study, every patient entered into the study should be assured of access to the best proven prophylactic, diagnostic or therapeutic methods identified by the study." Making a successful intervention available to participants after a trial is an important ethical obligation. There is however a related obligation to ensure that participants are no worse off during the trial than they were before it. In short, research participants should not be made worse off as a result of their inability to have continued access to the successful intervention after the trial has ended.

Once it is recognized that research projects should sometimes arrange to provide post-trial benefits to the participants, an ethical question arises about the justice of differentiating between former trial participants and others in the community in general who need similar medical treatments.

Thus a plan for the routine provision of a successful new intervention to participants after a trial has been completed is a way to ensure that the study is responsive to the health needs of the country. The ethical obligation to provide the intervention to others in the community who might benefit from it is considerably less strong, but a plan to do so would help reduce the risk of exploitation.

16.7 Need for Developed Countries to Assist Developing Countries in Capacity Building

Increasingly, scientists from developing countries are becoming more involved as collaborators in research, and many of them have developed their capacity for technical contributions to research projects and for appropriate ethical review of research protocols. Given the magnitude and severity of the health problems in different countries, capacity building to address ethical issues that arise out of collaborative research must be promoted on a priority basis.

An ethically relevant feature that distinguishes most developing countries from developed countries is the lack of access to adequate health care by a large majority of the population. Many developed countries have long provided universal access to primary health care through a national health service or government based insurance system. However, in the developing world, fewer health care services with limited access are available. Access to health care is an important issue in research ethics, because an ethically appropriate clinical trial design requires an assessment of the level and nature of care or treatment available outside the research context, as well as any possible future health benefit that might arise from research.

A unique feature of international collaborative research is the degree to which economically more prosperous countries can enhance and encourage further collaboration by leaving the host community or country better off as a result. The provision of post-trial benefits to participants or others in the form of effective interventions is one option. In most cases, offering assistance to help build local research capacity is another viable option. Making the necessary resources available for improving the technical capacity to conduct and sponsor research, as well as the ability to carry out prior ethics reviews, is one way to move forward in this effort.

In developing countries the regulatory processes are somewhat ill-defined. The weakness in the pharmacovigilance system may be attributed to the understaffed regulatory office and also to the lack of expertise to review and take necessary decisions. Further, the lack of awareness and training along with constraints on resources create problems for conducting GCP compliant trials in developing countries. Hence harmonization with guidelines and regulations is difficult.

Conclusion

Just as good science requires appropriate research design, consideration of statistical factors, and a plan for data analysis, it must also be based on sound ethical principles. Only then can research succeed in being efficient and cost-effective, while at the same time embodying appropriate protections for the rights and welfare of human participants. The unprecedented international movement of people, goods and ideas has made people in developed countries more aware of the imbalance in the global burden of disease. This is sometimes referred to as the "10/90 gap"- less than 10% of global health care expenditures are devoted to diseases that account for 90% of the global burden of disease. Clinical research that are responsive to the health care needs of these countries constitute one such initiative, but it is neither necessary nor desirable to relax our ethical standards in order to achieve this goal. On the contrary, the standards should be maintained, with circumscribed adaptation to the needs of developing countries. Thus in conclusion, the numerous examples presented have shown that a wide range of ethical issues may arise in any clinical study, which raise many complex questions which are subject to many interpretations and solutions.

Our greatest challenge is to realize and fully accept that in all research involving human subjects, ethics and science are not separable — a given study must conform to ethical standards or it should not be performed, and it must be scientifically sound or it cannot be ethical. The use of a good research design and adherence to sound ethical principles should result in the conduct of research that is valid, reliable, and ethically acceptable in any country. If investigators understand this seemingly simple concept, the ethical dilemmas will not vanish or suddenly become easier to resolve, but we will at least be less likely to conduct activities in the name of science that are disrespectful or harmful to others.

"The future direction of the pharmaceutical sciences must be shaped not only by science, but also by economic, political and ethical issues. The latter issues are likely to be more challenging than the science itself"

Dr.David J Triggle
Professor, SUNY,
Buffalo, USA

Suggested Reading

1. Greg Koski, Ph.D., M.D. ; Stuart L. Nightingale, M.D. Research Involving Human Subjects in Developing Countries . NEJM Vol 345; No 2; July 12,2001; pg 136-8.
2. Ezekiel J. Emanuel, M.D., Ph.D. Franklin G. Miller, Ph.D. The Ethics of Placebo-Controlled Trials-A Middle Ground; NEJM 20 Sept 2001; vol 345 no 12 pg 915-919.
3. Spilker B. Guide to clinical trials. New York 1991. Raven Press.
4. Ethical Guidelines for biomedical research in human subjects. ICMR publication.
5. Nuffield Council of Bioethics. The ethics of research related to health care in developing countries.

CHAPTER 17

Informed Consent in Clinical Research

P. Usha Rani and M.U.R. Naidu

17.1 Introduction

Ethical clinical trials are built on the foundation of informed consent, the process through which patients learn the intricacies of the research in which they are about to become involved. Consent is derived from Latin word "Consentire" which means "to agree". Informed consent is a process by which potential clinical trial participants voluntarily confirm their willingness to participate after being informed of aspects of the trial that are relevant to their decision to participate. Informed consent is documented by means of a written, signed, and dated informed consent form.

Informed consent applies to two different but related settings: medical practice and biomedical research on human tissues and health information. In the former, it is at the heart of the relationship between doctor and patient, however in the latter setting, it is medical research, where informed consent is mandatory before extracting individual's information or using their biological material like tissues or organs. Only after obtaining consent to their uses, can researchers proceed both ethically and legally.

Thus ***Informed consent*** *refers to* a process where by - Any individual (or their legally authorized representative) who is asked to take part in a research study should be able to understand as completely as possible what procedures, risks, benefits, alternatives, and rights are involved, and make the choice about participating in the study without pressure or undue inducement to participate. Further it is an ***Ongoing process,*** **which means -** Free and informed consent is not one event but an ongoing process. Throughout the study, participants must be told about changes in the study, feel free to ask questions, and be allowed to withdraw from the study at any time for any reason, or even for no reason

The key word here is "process," for in clinical trials, consent is more about education than a simple agreement to participate. Beginning during a trial's recruitment phase, and often lasting beyond the duration of the trial itself, informed consent is an ongoing discussion among patients, researchers and medical staff that encompasses procedures, possible risks and benefits and, in general, lays out the experience participants can expect during the clinical trial's course. Informed consent is the process of learning the key facts about a clinical trial before a person decides whether to participate, or whether to allow a child or incapable adult to participate. Although all research participants must sign a special form to proceed with the trial, **informed consent is much more than signing a form**. It is a process that continues throughout a study. Participants or their legal representatives should feel free to ask the research team questions before, during and after the study. Informed consent continues as long as someone is in the study.

Informed consent is about making sure participants are fully engaged. For patients, trials offer benefits and risks, advantages and inconveniences. For researchers, it's an article of faith that patients can't sincerely decide whether to participate until all of these aspects, the positives and the negatives, have been detailed in understandable terms. This means both written material and direct conversations must rely on lay language ie "plain English" – and not medical jargon or legalese. To be sure, when some prospective participants learn the particulars of a trial, they may decide not to proceed. This is as it should be. For researchers, recruiting participants based on anything less than full disclosure would call the entire effort into question. Informed voluntary consent underlies everything.

Patients enter clinical trials for a variety of reasons. Some want to receive the most cutting-edge treatment for a condition or disease. Others want to participate in a project that might aid a large number of people. Some are simply fascinated by research and want to be a part of it. Whatever their reasons, informed consent ensures patients' goals are in concert with a particular trial's purpose. Painstakingly, researchers describe every aspect of the effort, from potential harms and discomfort to the simple inconveniences involved with visiting an investigator on a regular basis or, in some cases, checking into a research facility for overnight stays. If during the course of a trial new risks are discovered, patients are informed promptly and consent materials are revised. It's a dynamic process that doesn't end simply because a form has been signed at the beginning of a trial.

17.2 Historical Background

History reveals that in past human subjects have been abused for "research" without their willing consent. The famous examples of such unethical practice are reflected in the Tuskeegee experiment and Experiments on German war soldiers.

Respect for persons requires that subjects, to the degree that they are capable, be given the opportunity to choose what shall or shall not happen to them. This opportunity is provided when adequate standards for informed consent are satisfied.

The Belmont Report
The National Commission for the Protection
of Human Subjects of Biomedical and Behavioral Research
April 18, 1979

Milestones in the development of informed consent include:

Nurember code - 1949

Helsinhki Doctrine – 1964

ICMR guidelines – 2000; 2002

DCGI Regulatory requirement – Schedule Y 2005

17.3 Elements of Informed Consent

Any trial related procedure will begin only after the written informed consent is obtained.

According to National guidelines – There are major and minor elements

Major elements of consent state important facts regarding the research activity. These facts include:

- Why the research is being done.
- What the researchers want to accomplish.
- What will be done during the trial and for how long.
- What risks and discomforts are involved in the trial.
- What benefits can reasonably be expected from the trial.
- What other treatments are available.
- The extent to which participant records will be kept confidential.

- The fact that participation is voluntary, and that a participant has the right to leave the trial at any time.
- Whom to contact if questions or problems arise during the course of a study.

There a few minor additional elements that must be present when appropriate. These include:

- A statement that the particular treatment or procedure may involve risks to the patient that are currently unforeseeable
- Anticipated circumstances under which the patient's participation may be terminated by the investigator
- Any additional costs to the patient resulting from participation in the research
- The consequences and procedures for withdrawing from the research
- A statement that significant new findings developed during the research will be provided to the patient
- The approximate number of patients who will be enrolled in the trial.

(Details of the same can be obtained from Schedule Y guidelines)

When a person is considering participating in a clinical trial, or giving permission for a child or incapable adult to participate, the research staff will provide information and informed consent documents that include the details about the study. They may also discuss the study with potential participants or their legal representatives. These documents have been reviewed by an Institutional Review Board, or IRB, the committee of scientists and non-scientists who must approve everything related to a clinical study before the study can begin or continue.

Documentation of informed consent: It is important requirement that informed consent be documented, in general, with a *written consent form* incorporating the required elements of informed consent. The participant or his/her legally authorized representative reads, signs and dates the consent form. The person obtaining consent also signs and dates the consent form. A signed copy of the consent form is given to the participant or, when appropriate, to his/her legally authorized representative. The whole process is documented in source notes along with the questions raised by the study participant and the explanations provided by the investigator.

17.4 Language Used in Consent Forms

Modern word processing systems can help generate language suitable for any level of education. The comprehension level of language used in informed consent should be tailored to the patient population. As there is no single universally accepted rule as to which style of writing is preferable, consent forms can be written in first, second or third person style or may also combine different writing styles.

For those people for whom English is not their native language, the consent documents will be provided in languages other than English. Since joining a clinical trial is an important decision, it is important that potential participants or their legal representatives should ask the research team any questions they may have about the study and the consent forms before making a decision and signing the form.

When possible, it is also acceptable to take the consent documents home and discuss them with family members or friends. Talking about the options can help potential participants feel comfortable with their decision.

If a person decides to join a clinical trial, or to give consent for another, he or she should be sure to ask for a copy of the informed consent documents so he or she can review them at any time.

17.5 Informed Consent Waiver

The IRB can waive the informed consent requirement only under specific, limited conditions. For example, informed consent can be waived, if (1) research procedures involve only minimal risks to human volunteers, (2) there is no jeopardy to their rights and welfare, and (3) the process of obtaining consent makes it impossible to conduct the research.

Minimal risk means - "the probability and magnitude of harm or discomfort anticipated in the research are not greater in and of themselves than those ordinarily encountered in daily life or during the performance of routine physical or psychological examinations or tests."

Waiving consent versus obtaining consent from a surrogate/legally authorized representative: When prospective research participants in minimal risk studies cannot consent for themselves because they are incapacitated in some way, the researcher must decide whether to request a waiver of all consent or use of a surrogate/legally authorized representative.

- Ethically, it is usually preferable to use a surrogate who knows the subject's wishes rather than waive consent entirely. However, these concerns are less serious for studies with minimal risks.
- If surrogates will be asked to give consent, even for studies involving minimal risk, the standard procedures for identifying a legally authorized representative and obtaining surrogate consent must be followed.

Consent for continuing or follow up procedures after initial waiver of consent: In occasional minimal risk studies, the consent may be waived when subjects initially cannot consent for themselves, but require consent for later procedures when subjects recover their capacity to consent.

Contents of follow up consent: The consent form should clearly describe what procedures occurred without the subject's consent, why they were performed, and what additionally will be done if consent is given to continue the subject in the study. The subject must also be given the option of refusing to allow the researchers to use the data already collected

17.6 Revising an Informed Consent

Situations may arise during the course of a clinical trial where informed consents need to be revised or new informed consents prepared. These situations include:

1. Data of interim analyses of an ongoing study demonstrates beneficial effect or increased of one treatment over the other
2. Occurrence of severe or serious adverse events that are probably drug induced
3. Sometimes sponsor may decide to discontinue further development of that particular medication.
4. Whenever there is a change in the investigator or major change in site facilities have occurred.
5. Any new issues which arise during the course of drug development and clinical trials since the day patient consented for the study.

17.7 Reconsenting Process

A trial participant may sometimes be required to reconsent in an ongoing trial. Situations where reconsenting may be necessary include:

1. Whenever the initial consent was not adequate
2. Any new safety information on the investigational product during the course of the study which may affect the subjects decision to continue in the trial.
3. Amendments in ongoing protocol which could influence patient continuation (e.g. biopsies, additional blood/body fluid samples, extension of study duration)

All revised consent forms and their translations require to be approved by the ethical review boards. Patients in an ongoing clinical trial are required to reconsent on the new revised informed consent form.

Evaluating and determining patient's understanding of the informed consents

It is important and often relevant to determine how well the patient has understood the contents of the informed consent. Approaches vary from simple verbal questions to extensive written tests to assess the degree of comprehension.

17.8 Consent Obtained from Special Groups of Patients

For many years, research with children and pregnant women was virtually nonexistent. People living in prisons and institutions, who could not fully participate in the consent process, were disproportionately represented. These vulnerable populations are now given special consideration and, in some cases, extra protection.

Pregnancy waiver

Until various terratogenicity and reproduction studies have been completed in the animals, women of child bearing potential should not receive any new investigational medicine. It is desirable that, female patients noy become pregnant while receiving an investigational medicine. A pregnancy waiver form is used because of this intension. This waiver may be considered as a separate informed consent or as an addendum to the informed consent. Patient of child bearing potential agrees either to abstain from sexual relationship during the clinical trial or to use contraceptive methods that have been approved.

Clinical trials in Children

Specific protection needs to be addressed when conducting clinical trials in pediatric population. The FDA's Pediatric Rule, demands that pediatric clinical trials be conducted for all new medications that will be – or could be – used to treat conditions or diseases in children. Some lessons learnt from studies that have already been done include : Medications need to be dispensed to children at far different doses than they once were. Researchers are learning that children's age can have a big effect on how they handle a drug. So can their gender, both because of differences in size and body composition and changes that accompany puberty in girls. Simply reducing the adult dose, therefore, is not a medically sound idea.

In such situations informed consent may be obtained from the parent, legal guardian or close relative. Children above the age of about 7 years should be asked for their own assent, in addition to obtaining consent from their parent(s) or guardian. There are models available for physicians to determine patients' competence like (a) communicating choices (b) understanding relevant information (c) appreciating current situations and (d) interpreting information rationally.

Questions Every Parent Should Ask About Pediatric Clinical Trials

- What are the IRB reviewers' comments and concerns?
- Is the study necessary? If the information can be gleaned from studies with adults or on animals, is there a need to put children at risk?
- What are the obligations and expectations for child volunteers? Are overnight stays required? Is a diary required? How many visits to the research center are required? What is the frequency of visits?

- What is the "escape rule"? That is, what adverse event or other type of discomfort has to happen for my child to be removed from the study? If that happens, will some alternative therapy be offered?
- Are there "stopping rules" regarding what might prematurely end the entire study? What safety-related events, or frequency of adverse events, would close the study?
- How much evidence is needed about an investigational drug's failure to work before a study is halted?
- If the experimental treatment is for a serious or life-threatening illness, will the study have some kind of special safety monitoring? Will study results periodically be reviewed by an independent third party?
- Are there any alternative clinical trials under way?

Special considerations for terminally ill participants

IEC / IRBs give special consideration to research involving terminally ill individuals, and often require special procedures for protecting the rights and well being of these participants. Special attention is paid to the informed consent process. As a general rule, the staff involved in these trials must give accurate information about eligibility for participation, treatment options, and risks and benefits, and communicate information about the trial in a manner that will not either give a potential participant false hope or eliminate all hope.

Special considerations for economically disadvantaged people

When reviewing applications for clinical trials that offer payment to participants, IEC / IRBs consider whether the amount or nature of any payment might be too much enticement, particularly if the subject population is economically disadvantaged.

Informed Consent for Genetic Research

For biogenetic research involving human participant, certain special considerations have to be borne in mind while obtaining informed consent of the participant enrolled in the study. These are in addition to general principles applicable to all medical research. Group consent will have to be taken from culturely appropriate authority like community head where there are no relevant authorities like village panchayat head, since genetic research gives rise to information applicable to community.from which the participants are drawn. The ethics committee should ensure that this has been applied where ever applicable.

Confidentiality and Privacy in Genetic Research

Confidentiality of subjects' study-related medical records is especially important in studies involving genetic research. Informed consent involves not only confidentiality of subject records but also privacy of subjects' identifiable health information [or "protected health information"], which is broader than records only. The informed consent regulations require that every Informed consent document have a statement describing the extent to which records that identify the subject will be kept confidential, noting that the FDA may inspect the records.

Defining risk and Benefits

Potential risks and benefits should be discussed thoroughly with prospective participants. Psychosocial rather than physical is the primary risk in genetic research, hence adequate counselling should be given to participants on the meaning of genetic information they receive. Only those persons qualified and experienced in communicating the meaning of genetic information should undertake genetic counselling..

Information Recommended for Inclusion in the Informed Consent Document for Studies Involving Genetic Research

- What specimens will be studied?
- How and where will specimens be obtained?
- Will specimens be stored for future, presently undefined study?
- Might these studies result in any commercially valuable products?
- Will samples be anonymous?
- How will confidentiality be maintained?
- What will be done with the information obtained on individual subjects?

  Will results be reported to the subject/parents/relatives?

  Will results be reported to the attending physician?

  Will results be recorded in patient's medical record?

  Will results be recorded in patient's medical record?
- Will there be prospective follow-up contact with the subjects?

Sample Consent Language Regarding Confidentiality

For genetic studies

Information identifying you will not be included with your blood sample for this research. The link between your identity and your sample will be destroyed before your genes are studied. Therefore, your identity will never be revealed to anyone or be connected with genetic information from your sample. Neither you, nor the study doctor will be able to know the results of any genetic analyses performed on your sample.

Challenges to Informed Consent

New developments in biomedical research and healthcare may mark the end of the traditional concept of informed consent. There are some shortcomings in the informed consent process while dealing with genetic disorders. As there is difficulty in comprehending some aspects, the concept of 'well informed' has become a challenge. Furthermore, the lack of training in genetics for physicians has added to the fallacy of the 'informational' component of consent.

A second challenge is the fact that genetic information often has an impact on people other than the patient, where as Traditional informed consent, however, assumes that any decisions that are made concern the values and life of a single individual. Further, traditional model fails to recognize that an individual is also a social person with a particular historical and socio-cultural values.

The latest and more serious challenge to informed consent, relates to the development of various DNA databases. Informed consent usually covers only specific and known uses of biological material—for example, when a person consents to their DNA being used in a particular experiment or research study. For other situations in which material may be used more than once, or for as yet unknown research, other forms of consent have been devised. Open, or blanket, consent is given only once, but covers any use of the material at any time in the future. This is particularly important for scientific research, in which new projects or experiments might be devised years after individuals have given their consent and deposited their biological material. Thus new emphasis on genetic predisposition to disease and prevention would therefore require an informed consent process that involves renewal and re-consultation, which is very much at issue with the developing population DNA databases.

In reviewing the new developments in genetic science and medicine, and particularly the development of population genetics databases, it seems logical that the old notion of informed consent has become outdated and needs review. It is also not clear yet whether the two main alternatives—presumed and open consent—will fare any better. Old rules often cannot fit new situations, and the changing needs, knowledge and globalization in biomedical and genetic research may demand a new ethical and legal framework for consent.

Checklist for Inspecting Informed Consent Process During Clinical Trial

S.No	Activities	Yes	No	N/A	Comments
1.	Does the title of the study appear at the top of the form				
2.	Does the name of the institute and investigator is present in the ICD (informed consent document)				
3.	ICD is on file for each patient enrolled.				
4.	Does ICD is translated into local language?				
5.	Does the back translation of ICD have been done?				
6.	Is the ICD has version number and date?				
7.	ICD version used is same to that approved by IEC.				
8.	Does consent/assent form clearly explain the procedures to be followed? (i.e. interviews, audio/ videotaping, medical procedures, physical exercises, experimental procedures, possible discomforts or inconveniences)				
9.	If risks are present, are procedures for reducing risks described?				
10.	If risks are present, is the medical liability statement appropriate?				
11.	If no risks are identified, is the standard risk statement included?				
12.	Are benefits of participation and other described? Otherwise, does the form state that there are no personal benefits?				
13.	Are confidentiality procedures fully described?				
14.	Does the data is stored in well controlled area where the entry for other personnel are inhibited?				
15.	How long the data will be kept?				
16.	How the data will be reported?				
17.	Does the consent/ assent form describe any compensation to be offered and detail alternatives if appropriate?				
18.	Have patient/witness signatures been personally dated?				
19.	Are there any discrepancies between dates the subject and the study representative signed the consent form?				
20	Has the signature/thumb impression of the patient been affixed with date?				
21	In the case of illiterate patients are there a signatures and details of impartial witness?				
22.	If the subject is incapable of giving informed consent (eg. Children, mentally ill, unconscious), has consent been obtained from a legally acceptable representative?				

S.No	Activities	Yes	No	N/A	Comments
23.	Does parental consent form have blank line for child's name?				
24.	Does mature minors and adolescents personally signed and dated on a separately designed written assent form?				
25.	Has informed consent been obtained before starting any study related procedures?				
26.	Did IEC approved study representative to obtain consent for all subjects?				
27.	If informed consent been administered by a designated person who is not medically qualified, is there evidence that subject's queries of a medical nature were answered by a medically qualified person or the investigator? Is the complete Informed consent is counter signed by investigator?				
28.	Are any risk associated to withdrawal made clear?				
29.	Are situations when participation can be terminated?				
30.	Does the consent/ assent form indicate the participant can withdrew at any time from the project with out reprisal or penalty?				
31.	Are situations when participation can be terminated described?				
32.	Are there signature columns for both subject and designee or researcher is clear?				
33.	Does the subject had and opportunity to take the consent home?				
34.	Is there documentation to supprot that subject's receive a copy of the signed and dated consent form?				
35.	Enough time is give to the subject for taking his/her decision?				
36.	Are there any changes/hand written corrections (e.g. contact telephone number) to any of the consent forms used?				
37.	Are yes/no or similar option on the consent form (if applicable)complete for all subjects?				
38.	Does consent forms are archived in an appropriate place?				
39.	Is the accessibility of data is restricted?				
40.	Is the complete address with telephone number of Principal investigator, person obtained informed consent and one representative from IEC is present.				
41.	Are original copies of IEC approved consent forms on file?				

Guidelines for preparing Inform Consent Form

1. Protocol Title with number, version and date. *Insert Title of Research Study*

2. Informed consent form language, number, version, date

Principal Investigator: *Insert Name of the Principal Investigator Address Insert Phone Numbers*

Emergency Contact: *Insert Emergency Contact Insert Phone Number/ Mobile number etc.*

*Answers to the following questions needs to be provided and explained in the subjects information part of Informed consent form.

Why am I being asked to volunteer?

- The subject is being invited to participate in a research study and why he/ she is being asked to volunteer
- Participation is voluntary
- The subject will get a copy of the consent form and should ask questions
- The subject will be asked to sign/put his/her thumb impression on this form if consent is given to participate
- Include some information about the study such as "a study of "X" (drug or device, etc.) in patients with "Y" disease.

You are being invited to participate in a research study. You are being asked to volunteer since you meet the requirements for enrollment into this study. Your participation is voluntary which means you can choose whether or not you want to participate. If you choose not to participate, there will be no loss of benefits to which you are otherwise entitled. Before you can make your decision, you will need to know what the study is about, the possible risks and benefits of being in this study, and what you will have to do in this study. The research team is going to talk to you about the research study, and they will give you this consent form to read. You may also decide to discuss it with your family, friends, or family doctor. You may find some of the medical language difficult to understand. Please ask the study doctor and/or the research team about this form. If you decide to participate, you will be asked to sign this form or put your thumb impression on the form.

What is the purpose of this research study?

- A concise explanation of the purpose of the research, incorporating any intent to assess safety +/- efficacy
- A Clarification that the drug/device is investigational. Can note that the drug/device is approved for another indication if applicable, but must clarify that the use of the drug/device in this study is experimental.

How long will I be in the study? How many other people will be in the study?

- Expected duration of a subject's involvement with the study
- Expected total duration of study.
- Total number of subjects in study.
- Include number of subjects at this site.

What am I being asked to do?

- A high level overview of the major elements of the study and what is expected of the subject (i.e. note here only the major procedures and milestones).
- Following the overview, provide a full list of procedures/tests by lay-term names, Consider including number of times each test will occur, amount, exposure if appropriate, etc. in easy layman terms.
- Describe each test/procedure in layman's terms
- (optional) May be complimented by a simple table or chart and/or other additional materials may be inserted here or given as a handout. Any such materials require IRB approval.

What are the possible risks or discomforts?

- Known risks from the study agent. May also be detailed in chart format and additional material inserted here or given as a handout. Any such materials require IRB approval.
- Risks, discomforts/inconveniences of study-related procedures noted in the section "What am I being asked to do?". If standard of care in testing is being changed, describe any resultant risk, if applicable. May also be detailed in chart format and additional material inserted here or given as a hand out. Any such materials require IRB approval.
- Clarify that if the subject is injured, they should inform treating physician that they are in a research study.
- Include information on reproductive issues, if appropriate. NOTE: If male contraception methods or warnings are warranted, the appropriate information must be provided in this section as well.
- Do not make statements of proven safety unless that safety data is part of Regulatory approved labeling. If the labeling safety data does not include data in the proposed study population for this study, make clear that there is no safety data in the population under study.

Reproductive risks: Due to the effect of this drug/device, there could be serious harm to unborn children (or children who are breast feeding) and it could also jeopardize the health of the mother. In addition, it is possible that harmful side effects that are not yet known could occur to both the mother and unborn or breastfeeding child. For this reason, if you are pregnant, you will inform the investigator and understand you will not be included in the study. If you are still capable of becoming pregnant, you will be given a serum pregnancy test prior to entry into the study. You also understand that you will practice a medically accepted method of birth control during your participation in the study. Further, you understand that while you are taking this drug/device you should not become pregnant, and if you do become pregnant, you must discontinue the drug/device and consider termination of the pregnancy (if required).

What if new information becomes available about the study?

During the course of this study, we may find more information that could be important to you. This includes information that, once learned, might cause you to change your mind about being in the study. We will notify you as soon as possible if such information becomes available.

What are the possible benefits of the study?

- A statement that there may not be any benefit (direct or indirect) to the subject
- Anticipated benefits that the subject may experience.
- Anticipated benefits to society

What other choices do I have if I do not participate?

- Information on other treatments available.
- Alternatives to entering the study including, when appropriate, supportive care with no additional disease-directed therapy
- A statement that they may discuss alternatives with their personal physician

Will I be paid for being in this study?

- Description of any monetary compensation (*payments/stipend), if subjects are being compensated for their time and travel.
- (optional) A breakup of the total compensation (i.e. clarify if paid after each visit/procedure or upon completion of the study, etc.)
- If there is no compensation for participation in this study, state that here

Will I have to pay for anything?

- Procedures or tests that will be covered by the study.
- Procedures or tests that are not covered by the study, stating how they will be paid for (i.e., third party payer (payor), etc.).

"No cost" language: You and/or your health insurance may be billed for the costs of medical care during this study if these expenses would have happened even if you were not in the study, or if your insurance agrees in advance to pay.

What happens if I am injured or hurt during the study?

- Provide contact information for research-related injury (i.e. can refer to the contact information noted in Consent header, if appropriate)
- Describe what treatment will be provided for research related injuries
- Explain how treatment for research related injuries would be paid
- Describe procedure for emergency care
- (optional) Subject's responsibilities relating to research related injuries

If you have a medical emergency during the study you may contact the Principal Investigator or Emergency contact listed on page one of this form. You may also contact your own doctor, or seek treatment outside of this Institute . Be sure to tell the doctor or his/her staff that you are in a research study being conducted at the Institute. Ask them to call the telephone numbers on the first page of this consent form for further instructions or information about your care.

In the event of any physical injury resulting from research procedures, medical treatment will be provided without cost to you, but financial compensation is not otherwise available from this institute. If you have an illness or injury during this research trial that is *not* directly related to your participation in this study, you and/or your insurance will be responsible for the cost of the medical care of that illness or injury.

When is the Study over? Can I leave the Study before it ends?

- Define when the overall study is to end.
- Explain what events could lead to early study closure
- Note that the subject can elect to leave the study at any time
- If early withdrawal could expose the subject to medical risks, describe how those risks will be minimized or prevented (e.g. in a hypertensive study, it may be necessary to wean a subject off the study medication or to transition them to alternate therapy).

This study is expected to end after all participants have completed all visits, and all information has been collected. This study may also be stopped at any time by your physician, the study Sponsor, the IEC or Regulatory Authorities without your consent because:

The Primary Investigator feels it is necessary for your health or safety. Such an action would not require your consent, but you will be informed if such a decision is made and the reason for this decision.

You have not followed study instructions.

The Sponsor, the study Principal Investigator, the IEC or Regulatory Authorities has decided to stop the study.

If you decide not to participate, you are free to leave the study at anytime without giving any reason for doing so. Withdrawal will not interfere with your future care.

Who can see or use my information? How will my personal information be protected?

The investigator and staff involved with the study will keep your personal health information collected for the study strictly confidential. Please refer to the separate "Confidentiality & Privacy Rights" document that explains more specifically how your personal information will be protected.

Who can I call about my rights as a research subject?

If you have questions regarding your participation in this research study or if you have any questions about your rights as a research subject don't hesitate to speak with the Principal Investigator or the Designee listed on page one of this form. Concerning your rights as a research subject, you may also contact the IEC member whose contact is mentioned in the form given to you.

When you sign this form or put your thumb impression, you are agreeing to take part in this research study. This means that you have read and /or understood the consent form, your questions have been answered, and you have decided to volunteer. Your consent also means that you are permitting this Organization to use your personal health information collected about you for research purposes within the institution and you are also allowing them to disclose that personal health information to outside organizations or people involved with the operations of this study.

A copy of this consent form will be given to you.

Documentation Of Informed Consent: Signature page:

Study code:

Protocol Title, number, version, date:

Subject's Initials:

Name:

Date of birth / Age:

I have read this study document and had its contents explained to me and understood the purpose of this study; what my participation in it will involve. I do freely give my consent to participate in this study, as described to me in this following document.

I voluntarily consent to participate in this study. I am aware that I may at any stage withdraw participation from the study without prejudice. (Subject to initial).

I authorize the release of my medical records to sponsor, ethics committee and Regulatory authorities. (Subject to initial ———).

I agree not to restrict the use of any data or result that arise from this study provided such a use is only for scientific purposes. (Subject to initial ———).

I understand that I will receive a signed copy of this consent form. (Subject to initial ———).

By signing this form I have not waived any of the legal rights which I otherwise would have as a participant in the research study. (Subject to initial ———).

__________________ __________________________

Name of Subject (Please Print) Signature Date

____

Thumb impression of Subject (If illiterate)

________ ______________________________________

* Name of Impartial witness/ Signature Date
Legally acceptable witness
(LAR) (Print Name)
(* If subject did not sign the consent)

Relationship of LAR to the study subject : ___________________________

____

Name of Person Obtaining Consent (Please Print) Signature Date

Sample Pregnancy Waiver

Only for female participants of child bearing potential

To all female trial participants of child bearing potential the following will be read and explained.

1. Based upon results of screening pregnancy test, trial participant is not known to be pregnant.
2. Pregnancy test will be conducted at specified intervals as per the study protocol.
3. Patients who are likely to become pregnant will voluntarily sign a statement indicating intent not to become pregnant during the course of trial and will agree to employ contraceptive methods acceptable to the investigator.
4. Investigational product administration will be immediately discontinued in case a participant becomes pregnant. Participant will be monitored by an obstetrician regularly throughout the pregnancy and by pediatrician after the child birth to assess the development of foetus and infant.

Voluntary statement of Intent to Avoid Pregnancy

I ____________________________________ understand the above statements, further I agree to attempt to avoid pregnancy while I am participating in this trial. Should I have sexual relationship during this period, I will employ a contraceptive method that has been approved by my physician. I will immediately contact

Dr ______________ if pregnancy is suspected. I am aware that I may decline to sign this statement and that my refusal to sign will have no effect on the further treatment of my disease.

Name of patient: _______________	Signature	Date:
Name of Investigator	Signature	Date
Name of witness	Signature	Date

Suggested Reading

1. Ethical guidelines for biomedical research on human participants. Indian Council of Medical Research. New Delhi. 2006. Chapter 6 pg 63-76.
2. Schedule Y – Amended version 2005.
3. Spilker B. Informed Consent and Review process. Chapter 27 pg. 202-218. In Guide to clinical trials. New York 1991. Raven Press.

CHAPTER 18

Monitoring in Clinical Trial

P. Usha Rani and M.U.R. Naidu

18.1 Introduction

Proper trial management and monitoring is necessary to assure adequate protection of the human subjects and the safety of all subjects involved in the clinical investigation in addition to maintaining quality and integrity of the resulting data obtained from the study. The clinical investigator is assisted in conducting the clinical trial by specialized professionals (study monitors, clinical research associates or "CRAs") working according to validated procedures and with the help of pre-established documents.

The purpose of trial monitoring as defined in ICH-GCP is to ensure that:

1. The rights and well-being of trial participants are protected;
2. Reported trial data are accurate, complete and verifiable from source documents and
3. Conduct of the trial is in compliance with the currently approved protocol/ amendments and applicable as per the regulatory requirements.

18.2 Types of Monitoring in Clinical Trials

Although a number of approaches are available to monitor trials, the appropriate procedure should be individualized based on the clinical trial requirements. One approach is by appointment of oversight committees including:

1. Trial management group – (TMG) - This should include individuals responsible for day –to-day management of the trial such as principal investigator, statistician, trial coordinator, research nurse, data management personnel etc.
2. Trial Steering Committee – (TSC) – Which provides overall supervision of the trial and ensures it is being conducted as per regulatory requirements and GCP. The committee monitors the progress of a trial including recruitment, data completeness and ensures there are no major deviations and violations.
3. Data Monitoring Committee – (DMC)- Reviews the accruing trial data and assesses whether there are any safety issues that should be brought to participants' attention. In the course of a double blinded trial, it is the only body that has access to unblinded data.

Thus sponsors of clinical investigations are required to monitor the progress of all studies. Considerable thought is required to monitor a clinical study. Most field monitoring requires regular visits to the site by sponsor's representatives or monitors.

18.3 Key Participant in Monitoring of Clinical Trial

Key participant in monitoring of clinical trial is the CRA and other members including the project manager, medical monitor, data manager and the statistician.

18.3.1 Sponsor's Monitor or Clinical Research Associate (CRA)

Interface with investigative site on behalf of sponsor or CRO. May be involved in any or all of the following:

- site evaluation and initiation
- assure regulatory requirements are met
- assure site personnel are trained and informed of the study protocol requirements
- assure site is provided with all materials required to conduct trial
- assure site personnel are qualified and aware of obligations
- monitor data and verify source records
- oversee the progress of the study at site level
- review the investigational products documents, utilization, storage and maintenance records
- may be an employee of sponsor, CRO or a private contract monitor

Project Manager-responsible for overall implementation of trial; oversee progress of trial; global trial communications; reporting to Regulators, budget and cost control.

Medical Monitor-individual with expertise in trial related therapeutic area; involved in protocol development, safety reporting, interpretation of data and clinical findings.

Data Manager -generate data queries to ensure quality of data; maintain ongoing data entry; monitor data analysis and provide reports.

Statistician-analysis of data – provide reports for data safety monitoring board (DSMB) or Endpoint committee

1. ***Selection of a Monitor:*** A sponsor may designate one or more appropriately trained and qualified persons (often referred to as "clinical research associates") to monitor the progress of a clinical investigation. The CRA must have thorough knowledge of good clinical practices, good human relations and good knowledge of the language of the country in which the study center to be monitored is located so that they can converse with the site staff effectively.

   A monitor need not be a person qualified to diagnose and treat the disease in question. Thus the number of monitors required for a study; as well as their education and training or expertise includes the following factors:

   - The number of investigators conducting the study
   - The number and location of the facilities in which the study is being conducted
   - The type of product involved in the study (eg. Medical device; drug for human use etc)
   - The complexity of the study
   - The nature of the disease or other condition under study
2. ***Written Monitoring Procedures:*** A sponsor should establish written procedures for monitoring clinical investigations to assure the quality of the study and to assure that each person involved in the monitoring process carries out their respective duties. Thus, standard written procedures (SOPs), may be used as a basic monitoring plan supplemented by additional project specific monitoring procedures.
3. ***Extent and Intensity of Monitoring:*** The extent and nature of monitoring needs to be defined prior to the start of the trial. The clinical trial risk assessment may be used to determine the intensity and focus of the monitoring activity, whilst the trial design would inform the methods used for monitoring. Further, the intensity of monitoring varies across studies and among sites. In early phase I studies, monitors may be required to be

present during all or part of a subject's treatment. The overall monitoring plan should remain fairly consistent, but the strategy for individual site may change during the course of the study depending on the study condition and site performance.

4. ***Frequency of monitoring visits:*** Sponsor is responsible for assuring that the investigator's obligations are being fulfilled and that the facilities used in the clinical investigation are acceptable throughout the study. Another major determinant in the monitoring plan is the frequency of the monitoring visits. The following are some of the factors which determine the frequency of monitoring visits:
   - Complexity of the protocol
   - Disease under study
   - Qualification and experience of the investigator and site staff
   - Number of subjects enrolled in the study
   - Rate of subject enrolment
   - Performance of the particular site
   - Number and location of the sites to be monitored by monitor
   - SOPs of the sponsor's monitoring
   - The monitor's experience and expertise

   The monitor should visit the investigator at the site of the investigation frequently enough to assure that:
   - The facilities used by the investigator continue to be acceptable for purpose of the study
   - The study protocol or investigational plan is being followed
   - Changes to the protocol have been approved by the IRB and/or reported to the sponsor and the IRB
   - Accurate, complete, and current records are being maintained
   - Accurate, complete and timely reports are being made to sponsor and IRB
   - The investigator is carrying out the activities agreed upon and has not delegated them to other previously unspecified staff.
5. ***Monitoring Activity:*** Clinical Research Associate's (CRA) **guide to monitoring clinical research** includes:
   1. Preinvestigation visits
   2. Training of investigator's and study personnel
   3. Preparing for a monitoring visit
   4. Monitoring informed consent documents

5. Monitoring subject recruitment, eligibility and records
6. Record of onsite visits
7. Conducting safety monitoring and monitoring of adverse events
8. Investigational medicinal products (IMP) storage, dispensing and accountability
9. Preparing for audits and detecting fraud
10. Prepare a Report on the each monitoring visit

Preinvestigation Visits: Sponsor through personal contact between monitor and individual investigator, is responsible for assuring that the investigator clearly understands and accepts the obligations incurred in undertaking a clinical investigation. Before initiating any study, the monitor should visit the site of the clinical investigation and assure that the investigator :

- Understands the investigational status of the test medication and the accountability requirement for the same
- Understands the protocol and the study plan in detail
- Understands the investigator's obligation in obtaining informed consent
- Understands the investigator's obligation to obtain IRB review and approval of the study protocol before the investigation is initiated and communicate the same with the sponsor
- Has suitable number of study participants for conducting the clinical trial
- Has the required facilities for conducting the study
- Has sufficient time from the routine obligations to carryout study related activities

Training of investigator's and study personnel ; During the launch of the study meeting, the monitor makes a short presentation of the important aspects of the protocol. Trains the investigating team on:

- How to fill the CRF
- How of dispense study medication
- Drug accountability
- Filling up of the various study specific logs
- ADR and SAE reporting
- Specific training in case of Web based CRFs and / or Drug dispensing system
- Breaking of the blinding in case of blinded study
- Handling biological samples
- Study specific documentation
- Any specific study related procedures as requested by the investigator

Preparing for a monitoring visit: Once the study is launched, the monitor should maintain regular contact with the investigating team. Monitoring visits may be planned at the beginning of the study as per the sponsor's SOPs and also as and when required during the study. Monitor must first inform the investigator about the visit and the proposed activity during the visit.

Informed Consent: Monitor should assure that the consent procedures result in a freely given and appropriately informed consent. Some clinical trials involve complex interventions. In such situations training may be imparted to the investigating staff so that the trial is presented in a clear, comprehensive and balanced manner. This technique may reveal deficiencies in the level of understanding, style of presentation, or extent of discussion in the consent process. While in other trials, it may be enough to simply check whether the ethical committee approved (translation, date, number and version) consent form has been signed and dated and that there is a record of the information discussed with the participants during the consent process.

Monitoring subject recruitment, eligibility and records: Compliance with source documents is an important integral part during monitoring of a clinical study. The following are some of the aspects to be checked during monitoring:

- Existence of patient from hospital records
- Informed consent documents
- Examination and visit dates
- Inclusion and Exclusion criteria
- Important laboratory test and other test results
- Concomitant medication
- Important Medical and Surgical history
- Adverse events and serious adverse reactions
- Withdrawal from therapy and /or followup
- Update the site master file
- Verify ethics committee documents / permissions/correspondence
- Drug accountability
- Study logs

It may not be possible to do 100% source data verification (SDV) with CRF and patient records. In some cases it may be wise to do 100% SDV in 20% of CRF. Alternatively, it may be wise to do 100% SDV for first level data like Informed consent forms; Inclusion criteria etc; 50% SDV for second line data like concomitant medication, results of important investigations etc and finally 20% SDV for third line data like concomitant disorders, medical history etc.

Record of onsite visits: The monitor makes a note of all the onsite activities. The persons present and the aspects discussed. Records whether all events are documents in the correct manner, compliance with source documents and accountability of study drug regimens. Any onsite corrections made in the CRF and whether it was done by the authorized persons. The onsite record, forms the basis of the monitoring report.

Conducting safety monitoring and monitoring of adverse events ; Monitoring safety parameters is an important monitoring aspect. It is wise to do 100% SDV for subjects with serious and/or severe adverse events. Monitor should conduct thorough monitoring of adverse events and safety lab parameters. Discuss with the investigator for any clarification on any safety issues.

Investigational medicinal products (IMP) storage, dispensing and accountability : One of the most important responsibilities of the monitor is checking the drug accountability and reconciliation. It is the monitor's duty to check whether the drug is stored securely, at the right temperature and all the logs are maintained by the designated persons. Further, only authorized members must have access to the investigational products.

Preparing for audits: It is monitors responsibility to help the investigating site to prepare for an Audit.

Prepare a Report on each monitoring visit: Monitors must complete a written report of each visit for the sponsor's files. The purpose of this report is to notify the project management team of the activities conducted, progress made, difficulties encountered. Every monitoring visit is followed by a monitoring report by the CRA which specifies aspects like:

- Date of visit
- Documents verified
- Aspects discussed with the principal investigator and study team
- Drug supplies present and further requirements
- Laboratory records/ reports/supplies
- Any onsite corrections made
- Any trial related aspects and/or problems encountered by the team

The reports are often in check-box format, with space for comments at the end. Site visit reports are legal documents that provide the Regulators with verification of a sponsor's diligence in monitoring investigative trials. Hence, monitors must seriously and carefully word the reports.

6. Closing Visit

Once the study is completed or terminated, the closeout visit is performed. During this visit, several important operations are completed which include:

1. collection of any remaining supplies related to study – Supplies of study medication which is used or that returned by patients, extra copies of protocol, investigator's brochure, unused case record form etc are recovered.
2. corrections – A session for corrections in documents may also be held during the visit. The site master file is completed and all records are verified and updated.
3. discussion with site team on study followup issues – Importantly, the investigator's role in preparing the report and signing it, later publications and the requirement to archive all study related documents and material for the legally – required period of time. Finally, a final visit report is written, with a quantitative evaluation of documents and supplies checked and left on-site or recovered.

18.3.2 Clinical Research Managers (CRM)

Most of the duties and responsibilities for monitoring the trials will fall upon the shoulders of the medical monitors and CRMs. Main roles of CRM are:

1. Preliminary visits to individual sites to ensure the smooth flow of data
2. Monitoring the progress of patient recruiting
3. Monitor and correct problems at investigator's site especially those related to: - Drug/device delivery
    - Data entry
    - Transfer of data from the investigator to central storage
    - Procedures for treatment and observation
    - Availability of manuals and forms
4. Maintaining physicians interest throughout lengthy trials
5. Conduct follow up visits at sites along with medical monitors so as to maintain their morale and to forestall problems with patients, the investigator, and the site staff
6. Verify signed printed case report forms and see that the investigator is timely paid
7. Working with statistician to develop interim reports and review the same in a timely manner.
8. Intractable problems may be referred to project managers or other higher officials.

Being a good monitor means, in large part, being a good manager. This involves planning ahead in case problems arise, developing strong communication with research personnel, and observing everything that occurs at a research center. To manage site related issues, monitors must possess good observational skills and be continually aware of all facets of their sites.

Suggested Reading

1. Bert Spilker. Guide to clinical trials. Raven Press 1999
2. Vogelson. The art and science of monitoring. Modern Drug Discovery 2001; 4 (3): 25-26
3. Rico-Villademoros et al. The role of the clinical research coordinator- data manager- in oncology clinical trials. BMC Med Res Methodol 2004; 4 : 6.

Check List for Monitors

I. Prior to Commencing the Study

S.No	Monitoring Activities	Yes	No	N/A
1.	Qualifications of the principal and co-investigator(s), i.e. written curriculum vitae			
2.	Ascertain principal and co-investigator's understanding of the data on the investigational product, including pharmaceutical, pre-clinical and if appropriate, clinical data (investigator's brochure);			
3.	The investigator's awareness of their obligations in undertaking the study and to conduct the same in accordance with the protocol, Declaration of Helsinki, SOPs, relevant legislation and good clinical research practice			
4.	The investigator accepts the relevant controls, including data verification procedures, audit and/or inspection;			
5.	Investigator has the means necessary to carry out the study safely with regard to factors such as availability of participants for recruitment, facilities, medical, paramedical staff and clerical support			
6.	Checks that Storage, dispensing and documentation of the supply of investigational product(s) is safe and appropriate and in accordance with local regulations and SOPs			
7.	Confirm that the regulatory requirements, ethics committee review procedures and informed consent procedures are followed and recorded			
8.	Discuss in detail the protocol and protocol related documents prior to obtaining the principal investigator's signed approval			
9.	Check the descriptions of the methods, normal and reference values for the tests to be performed during the study			

S.No	Monitoring Activities	Yes	No	N/A
10.	Provide the principal investigator with written information suitable for developing the patient information package			
11.	Obtain a copy of the relevant approval(s) from the local ethics committee and, where applicable, clinical institution and drug regulators prior to study commencement			
12.	Obtaining a signed and dated list of names of the staff directly involved in the study and a description of the functions of each person in the study			

II. During The Course of the Study

S.No	Monitoring Activities to Ensure Yes	No	N/A	
1.	Each investigator has access to a mechanism to identify the treatment of a particular participant			
2.	Informed consent for each participant has been obtained prior to commencement of any protocol required activity for that participant			
3.	All investigational products are used according to the conditions in the protocol			
4.	Study participant compliance is properly documented			
5.	Source data are available, to ascertain the existence of the participant and enable information recorded in the CRFs, e.g. biological results, radiographs, to be verified where possible. The source data should also be checked for correct labelling, dating, signatures, etc. (Source data may be the original or a signed and verified copy, as agreed with the sponsor). Access to source data must take place within the requirements of the privacy legislation and is dependant on the express informed consent of the participant which may be collected at the time of original enrolment into the study			
6.	Any changes to the CRFs are properly documented, signed and dated			
7.	Complete dispensing or distribution records of the investigational product(s) are secured and maintained.			

Contd...

S.No	Monitoring Activities to Ensure	Yes	No	N/A
8.	The progress of the study, determination of recruitment status, number of withdrawals and adverse events documented			
9.	Monitor should make an effort to obtain maximum information on any missing or unclear data			
10.	Conformity of the data presented in the CRFs with source data			
11.	Essential documents are being correctly filled in and stored in compliance with the protocol			
12.	Documentation of informed consent procedure			

III. After Completion of the Study

S.No	Monitoring Activities to Ensure	Yes	No	N/A
1.	Complete documentation of all clinical and laboratory investigations is available in the CRFs			
2.	All CRFs are collected and placed for safe keeping in accordance with regulatory requirements, privacy legislation and the sponsor's operating procedures			
3.	The principal investigator has notified the study participants, ethics committees and, if required, regulatory authorities that the study has been completed			
4.	Treatment codes are collected, recording all cases where the treatment code has been broken and the reasons for doing so			
5.	The principal investigator is aware of the archiving requirements for source data and primary records			
6.	The standard operating procedure for unused supplies of the investigational product(s) is followed, i.e. unused supplies are collected or destroyed; that an appropriate record is kept in collaboration with the pharmacist or person designated as responsible for handling the investigational product(s			
7.	All queries resolved before data base is locked			
8.	The study information is collated and sent to the reporting centre for multi-national studies			

Flow Chart – Start of Study

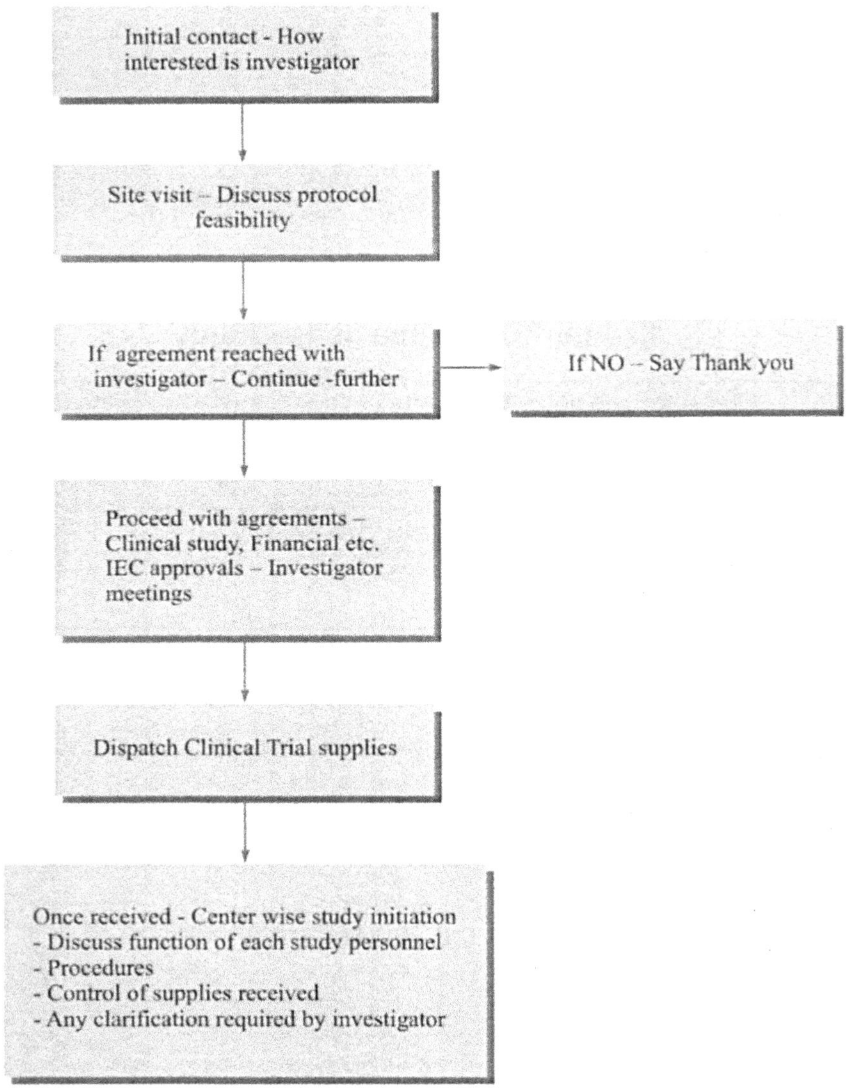

CHAPTER 19

Quality Assurance in Clinical Research

P. Usha Rani and M.U.R. Naidu

Quality Assurance

Quality assurance in principle means, "All clinical research activities are established to ensure that the trial is performed and the data are generated, documented (recorded), and reported in compliance with GCP guidelines and the applicable national regulatory requirements."

Quality control (QC): periodic operational checks within each functional department to verify that clinical data are generated, collected, handled, analyzed, and reported according to protocol, SOPs, and GCPs.

Quality control (QC): periodic operational checks within each functional department to verify that clinical data are generated, collected, handled, analyzed, and reported according to protocol, SOPs, and GCPsQuality control (QC): periodic operational checks within each functional department to verify that clinical data are generated, collected, handled, analyzed, and reported according to protocol, SOPs, and GCPsQuality control (QC): periodic operational checks within each functional department to verify that clinical data are generated, collected, handled, analyzed, and reported according to protocol, SOPs, and GCPs.

Operational Plan

The operational quality control and assurance plan describes how the quality processes will be applied throughout the clinical trial. It defines the various quality-related tasks in the study. A quality plan documents specific quality practices, resources, and activities relevant to a specific project.

In fact, complete QC plan for assessing the overall project management addresses the following:

- investigator selection and qualifications
  - experience in conducting clinical trials
  - experience with the specific indication
  - not on the FDA's restricted or debarred lists
  - adequate staff and facilities
  - personal involvement
- study conduct (monitoring)
  - subject informed (signed informed consent form)
  - subject's eligibility (inclusion/exclusion)
  - protocol compliance
  - adverse events (AEs) and concomitant medication
  - drug accountability and storage
- source document verification
  - medical records
  - lab data
  - progress notes
  - diagnostic tests
- query resolution
  - completed data clarification forms
- compliance with regulations

Site Management Metrics

19.1 Monitoring-Details Presented in Chapter on Monitoring

19.2 Audit

A sponsors perform audits, as part of implementing quality assurance, where the activities of the sponsor and / or the activities of the investigators and CRO may be audited. The Audit Trail refers to the ability to track the flow of data retrospectively as it moved on from the patient through lab samples and labs report, through the steps of data processing, analysis, documentation and the final report generation. Thus the audit trail consists of documentation of what occurred at each step of the process, particularly if the data are transformed, corrected or modified in any way. Thus an audit trail ascertains the accuracy of data as it moved from original form to final report.

19.2.1 Purpose

The purpose of a sponsor's audit, which is independent of and separate from routine monitoring or quality control functions, is :

I. To evaluate trial conduct and compliance with the protocol, SOPs, GCP, Good Laboratory Practices (GLP - where appropriate) and the applicable regulatory requirements and

II. To ensure that Clinical monitors, investigator's and investigating staff are doing their job appropriately

III. Data is suitable for regulatory submission and ensure future regulatory audits will be uneventful.

19.2.2 Selection and Qualifications

The sponsor should appoint individuals, who are independent of the clinical study /systems, to conduct audits. The sponsor should ensure that the auditors are qualified by training and experience to conduct audits properly. An auditor's qualifications should be documented.

19.2.3 Auditing Procedures

The sponsor should ensure that the auditing of clinical study/systems is conducted in accordance with the sponsor's written procedures on what to audit, how to audit, the frequency of audits, and the form and content of audit reports. The sponsor's audit plan and procedures for a trial audit should be guided by the importance of the trial to submissions to regulatory authorities, the number of participants in the trial, the type and complexity of the trial, the level of risks to the trial participants, and any identified problem(s).

The observations and findings of the auditor(s) should be documented and accessible to the ethics committee and / or the regulatory authority. When required by applicable law or regulation, the sponsor should provide an audit certificate. However, a sponsor's audit is for internal use and therefore a copy is not usually given to the investigator audited.

Audit may be (I) Systems Evaluation or (II) Data verification and validation.

Systems evaluation includes the number of monitoring visits along with documentation of each visit; prestudy and protocol review meetings; reporting of serious adverse reaction to regulators and investigators; documentation of all study related correspondence especially with investigators and also calibration and validation of computer programmes if the same are used for data capture in clinical trials. Finally, verification of storage facility for study documents and other clinical trial supplies.

Data verification and Validation emphasizes the integrity of the data to the sponsor and regulators. Major areas of interest to an auditor are documentation of informed consent, patient selection, enrolment and degree of patient compliance and serious adverse events if any reported. All essential documents including ethics committee submission letters, ethics committee approval letter, trial master file with all required documents will be verified.

The audit may be either in-house evaluation of all clinical trial data; on-site audits; analysis of information and / or reporting of results.

19.2.4 Non-compliance

Noncompliance with the protocol, SOPs, GCP, GLP and/or applicable regulatory requirement(s) by an investigator, or by member(s) of the sponsor's staff should lead to prompt action by the sponsor to secure compliance. If the monitoring and/or auditing identifies serious and/or persistent noncompliance on the part of an investigator, the sponsor should terminate the investigator's participation in the trial. When an investigator's participation is terminated because of noncompliance, the sponsor should promptly notify the regulatory authority(ies).

19.3 Inspections

An regulatory inspection is not an indictment and generally does not mean that the investigator's work is suspect. It's simply a quality assurance process used to verify clinical data and regulatory compliance.

Types of Inspections

The most common type of inspection is classified by the FDA as a "Routine" inspection and is generally triggered by a New Drug Application (NDA) submission. Hence clinical investigators who enroll the most patients in the NDA's pivotal trials are the most likely candidates for a routine inspection.

"For Cause" inspections are much more infrequent and generally arise only when the agency receives reports of, or otherwise becomes aware of, suspicious behavior by a clinical investigator. conducting a large volume of clinical trials;

Following are a list of reasons which may call for a regulatory "For Cause" inspections;

- conducting clinical studies outside of one's field of specialization;
- reporting significantly better efficacy, fewer adverse effects, or different laboratory results than other investigators studying the same drug;
- having apparent access to too many patients with a specific disease state for the locale or practice setting;
- complaints from a patient or sponsor of an alleged violations of the regulations, protocol, or human rights.

Prior to the inspection, the following study related documents only must be assembled in a secure place:

- Protocol
- Investigator's Brochure and IND Safety Reports
- Form FDA 1572 with accompanying CVs
- IRB correspondence, including approval documentation and final report to IRB and Sponsor
- IRB-approved Informed Consent form
- IRB-approved advertising
- Study-related correspondence, excluding investigator agreement and financial information
- Monitor sign-in Log
- Laboratory certification documents
- Drug accountability records
- Each subject's signed informed consent
- Also, assess the support areas (pharmacy, lab) must be properly prepared; incase the regulators tour the facility.
- Finally, be prepared to answer the following questions:
- Where was the study done?
- What special equipment was used?
- Who assisted in performing the study?
- What were each person's specific duties?
- Describe the sponsor's monitoring procedures and your interaction with the monitor.
- How did you account for the drug received, dispensed to/returned from subjects? Was all drug returned to the sponsor?

The regulatory authority may inspect the conduct of a clinical trial by on-site visits. Inspectors may conduct both announced and unannounced inspections especially if there is suspicion on the competency of the investigating site or evidence of GCP non-compliance. Such an inspection which may be carried out at random and / or for specific reasons, should consist of a comparison of the procedures and practices of the clinical investigator with the commitments set out in the protocol and reports submitted to the regulatory authority by the investigator or the sponsor.

Further, the facilities and SOPs of archives should be reviewed for retrieval. Inspections may include a data audit. The regulatory authority should have easy access to all patient files and raw data utilized for and generated during the trial. All site data and documents must be available for verification. All observations and findings should be verifiable in order to ensure the credibility of data and to assure that the conclusions presented are derived correctly from the raw data. Verification processes must, therefore, be specified and justified.

The following items are routinely examined:

- adequacy of communication with the IRB, including the initial submission document, adverse event reporting, and progress reports;
- completeness of accountability documentation for the receipt, storage, administration and return of test article (drug, device, etc.);
- compliance with the study protocol and documentation that each deviation/amendment received IRB and sponsor approval;
- appropriateness of the informed consent process (Was the patient was properly consented? Was the correct IRB-approved version used?);
- prompt and complete reporting of adverse events to the IRB and sponsor;
- compliance with the record retention requirements and that the investigator had immediate access to the study records during the trial; and
- adequate monitoring of the site and communication with the sponsor.

19.4 Data Management Metrics

Examples of data management metrics for QA are:

- percentage of database errors
- percentage of queries manually generated
- time from last patient out to database lock
- number of times a locked database is opened.

Summary

Management of clinical research data involves:

- Management of compliance with the protocol, SOPs, and GCPs
- Help to resolve systemic problems before the end of the study and thus reduce generation of data query
- Maintenance of data integrity throughout the study's course and that the data collected are the data required by the protocol
- Maintenance of accuracy and consistency of data from entry into the CRF to final datasets reported in the final CSR

Suggested Reading

1. Bert Spilker . Guide to clinical trials. New York 1991. Raven Press.
2. Surviving an FDA Inspection of Your Clinical Trial Site – Emissary International Publication-2008
3. Nickols Pc. Principles of clinical research. Wrightson Biomedical Publishing Ltd. 2001.
4. Graham D Ogg. A practical guide to quality management in clinical trial research. Taylor and Francis publication.

Check List for Clinical Trial Protocol and Ethical Issues for Evaluation by Auditor

1.1 Previous research relating to safety and potential benefit of intervention

- Do the results of laboratory and animal studies provide sufficient indication of the potential benefits and safety of the trial treatments in humans?
- Has appropriate information been supplied regarding the kinetics and dynamics of the trial treatments?
- Is this trial necessary? Do the investigators refer to a rigorous, preferably systematic review, of all previous trials that show the proposed trial would contribute further to existing knowledge?

1.2 Trial Methods

- Are the objectives of the trial clearly defined ?
- Are the selection criteria for entry to the trial appropriate?
- Is the method of enrolment of study participants defined ?
- Are the treatments well defined?
- Is the sample size justified?
- Is the randomization procedure described?
- How is the blinding maintained and allocation to treatment groups be concealed?
- How are the outcome measures assessed ?
- Will both benefits and harms of treatment be assessed?
- Will both prognostic factors be considered?
- Is the duration of post-treatment follow-up stated?

1.3 Ethical Issues

- Will the trial be supervised by a clinically competent, medically qualified person?
- Have the specific roles of each investigator been stated? Are these roles commensurate with the qualifications and experience of the investigators?
- Does the investigator plan to obtain informed consent (in writing if possible)

from study participants? Will all potential participants be adequately informed about the aims and methods of the study; the anticipated benefits, potential hazards and inconveniences associated with being in the study; and their rights to withdrawal from the study without prejudicing further treatment?

- Is the process of informed consent documented clearly in source notes?
- Has the trial protocol been approved by an appropriate ethics committee of the participating institution?
- Has the remuneration to be received by the investigators and /or participants been disclosed?
- Have suitable arrangements been made to monitor protocol compliance, drug dispensing, adverse effects and data processing?
- Have suitable arrangements been made for interim analyses and how will stopping rules be applied?
- Is there a clear statement confirming that the applicant has satisfactorily addressed insurance and indemnity issues?
- Is the subject informed of what methods of blinding and treatment allocation and post study treatment?
- Will the results of the study be revealed to the subject after report is compiled?

Regulatory Audit Checklist

I. The following documents need to be organised

1. The Regulatory Binder arranged by general heading
   (a) Protocol, include all versions
   (b) Investigator's Brochure, all versions
   (c) Informed Consent Form, all versions
   (d) Protocol Amendments
   (e) Form FDA 1572, all versions
   (f) CVs for PI and Subinvestigators listed on all versions of form FDA 1572
2. All communications to be filed in chronological order
   (a) Correspondence with Sponsor/CRO
   (b) Monitors reports and monitoring Log
3. Communications with IRB / IEC
   (a) Date of IRB notification and date of IRB acknowledgement and / or approval
   (b) Approval Letter (initial) for initial protocol with original informed consent

(c) Amendment approval(s) with the approved informed consent
(d) Periodic updates or Annual Reports
(e) Renewal Documents
(f) Notification of: Serious Adverse Events
(g) Acknowledgement of: IND Safety Reports, Study Termination, Final Summary

4. Laboratory Related Documents:
   (a) Laboratory Certification and normal ranges
   (b) CV of laboratory director
5. Medicinal Products Accountability and Drug Logs
   (a) Receipt of Drug
   (b) Dispensing
   (c) Return
6. Clinical Research Participant Related Documents
   (a) Informed Consents for screened / enrolled subjects
   (b) Source documents for each subject enrolled (including labs, x rays, scans, etc.)

II. Review the Following

1. For each subject enrolled review
   (a) Review Inclusion / Exclusion Criteria
   (b) Document reason for excluded subjects
   (c) CRFs completed for each enrolled subject
   (d) Source documentation for all CRF entries
   (e) Data Clarification issues satisfied
   (f) Consent obtained for all subjects screened / enrolled
   (g) Verify correct version of informed consent signed
   (h) Confirm 'Notes to File' present as appropriate
2. Medical Records and / or Study Files
   (a) Condition of subject at time of entry into the study
   (i.e., all inclusion / exclusion criteria met)
   (b) Exposure to Study Drugs
   (c) Concomitant medications
   (d) Laboratory reports
   (e) Diagnostic tests
   (f) Dose Modifications and concomitant medications
   (g) Adverse events – start date and stop dates

Clinical Trial Audit Checklist – Few Important Issues

No.	Audit Element	Yes	No	NA	Notes
Does the IC Form have all essential elements enlisted in Schedule Y ?					
	Statement that the study involves research				
	Expected duration of subject's participation				
	Description of procedures to be followed, including all invasive procedure				
	Description of any reasonably foreseeable risk or discomforts to the subject				
	Description of any benefit to the subject other reasonably expected from research				
	Disclosure of specific alternative procedure/ therapies available				
	Statement describing the extent to which confidentiality of records identifying the subject will be maintained				
	Trial treatment schedule and the probability for random assignment to each treatment				
	Compensation/or treatment available to the subject in the event of trial related injury				
	Contact details about whom to contact for trial related queries, rights the subjects				
	Anticipated prorated payments				
	Subjects responsibilities in participation				
	Statement that the participation is voluntary				
	Have there been any serious adverse events (SAEs)?				
If Yes					
	Have Serious adverse event forms been filled for all the SAEs?				
	Have all SAEs been reported to the Sponsor within 24 hrs of occurance?				
	Has any patient been withdrawn due to an Adverse event (AE)? If so, is the study termination form filled correctly?				
	Have patients with AE been followed up until resolution, stabilization or until another cause for AE determined?				
	Has adequate medical care been given to the patient especially in the event of intercurrent illness, adverse event including abnormal lab parameters?				

No.	Audit Element	Yes	No	NA	Notes
Does the IC Form have all essential elements enlisted in Schedule Y ?					
	IEC/IRB approvals- Do the IRB/IEC approval letters state the following?				
	Study Code, Title and Version number of the protocol				
	Name and address of IRB/IEC/EC indicated in the approval letter and compare it with the one stated in investigator's undertaking				
	List of other documents reviewed				
	List of members present at the meeting (with their designations)- Is the quorum of the five members identified in Sch Y satisfied?				
	Date, time, venue of the meeting identified in the letter				
	Signature and date of Chaiman/Member secretary				
	IRB/IEC Approvals as above for amendments?				
	IRB/IEC Approvals as above for promotional materials (advertisements etc) used for patient recruitment				
	Has the investigator submitted reports of all SAEs to the IRB/IEC?				
	Has the investigator apprised the EC/IRB/IEC about the trial progress?				
	Regulatory approval letter				
	Is DCGI permission for conduct of clinical trial available?				
	If yes, does the date of approval precede the date of trial initiation?				
	Protocol				
	Has the right version of the protocol been used?				
	Has the clinical trial protocol been reviewed and signed off according?				
	Investigator's Brochure (IB)				
	Is the latest version of the IB available at the site?				

No.	Audit Element	Yes	No	NA	Notes
Does the IC Form have all essential elements enlisted in Schedule Y ?					
	Are the appropriate reference ranges for medical/laboratory/technical procedure(s) and/or test(s) filed in Investigator's Trial Master File?				
	If there are any changes in reference ranges, has the investigator informed the changes to the Sponsor?				
Laboratory reference ranges					
	Are all the entries in CRFs legible, completed properly and made in the appropriate fields?				
	Have CRF data corrections been signed and dated by the Investigator/authorized site staff?				
	Have the original entries been scored off in way that the original entries are still visible?				
	Where needed, have reasons for the changebeen mentioned?				
	Does each subject meet all eligibility criteria?				
	Have Efficacy assessments been made as per protocol?				
	Are the patient visit dates according to protocol?				
	Have patient compliance data been entered?				
	Have results of laboratory and other investigations been transcribed correctly on to CRFs?				
	Has the clinical significance of abnormal lab values been commented upon?				
	Have lab abnormalities been followed up?				
	Determine whether all adverse events reported in the CRFs				
	Determine whether concomitant therapy indicated in the CRF where applicable?				
	Handling of Clinical Supplies				
	Are adequate records of receipt of clinical supplies maintained?				
	Are the Drugs being used for protocol purpose only?				

No.	Audit Element	Yes	No	NA	Notes
Does the IC Form have all essential elements enlisted in Schedule Y ?					
	Are the emergency code break envelops securely stored?				
	Are the drug accountability records correct?				
	Are the storage conditions/monitoring methods (if any) as per protocol?				
	Have errors in dispensing or administration of drugs been documented?				
	Is the trial medication being maintained under controlled access?				
	Have un-used trial medications been returned to the sponsor or disposed of according to protocol?				
	Are the drugs dispensing records being maintained properly?				
	Maintenance of Source Data				
	Are all source documents available at the site?				
	Are subject ID/Name/Hospital registration number available on each record?				
	Have source data been correctly transcribed on to CRFs?				
	Is the source data maintained properly and kept securely?				
	Have thermal papers carrying data been photocopied and filed?				
	Records Retention				
	Is adequate space available at the site for retention of documents?				

CHAPTER 20

Pharmacovigilance and Drug Safety

P. Usha Rani and M.U.R. Naidu

20.1 Introduction

Pharmacovigilance (PV) is the pharmacological science relating to the detection, assessment, understanding and prevention of adverse effects, particularly long term and short term side effects of medicines. Generally speaking, pharmacovigilance is the science of collecting, monitoring, researching, assessing and evaluating information from healthcare providers and patients on the adverse effects of medications, biological products, herbalism and traditional medicines with a view to:

- identifying new information about hazards associated with medicines
- preventing harm to patients.

The term pharmacovigilance is derived from - pharmakon (Greek), "drug;" and vigilare (Latin), "to keep awake or alert, to keep watch."

Pharmacovigilance is particularly concerned with adverse drug reactions, or ADRs, which are officially described as: "A response to a drug which is noxious and unintended, and which occurs at doses normally used... for the prophylaxis, diagnosis or therapy of disease, or for the modification of physiological function.

Pharmacovigilance is the science and activities relating to the detection, assessment, understanding and prevention of adverse effects or any other possible drug-related problems. Recently the also includes herbals, traditional and complementary medicines, blood products, biologicals, medical devices and vaccines.

20.2 Terms Commonly Used in Drug Safety

- *Benefits* are commonly expressed as the proven therapeutic good of a product, but should also include the patient's subjective assessment of its effects.
- *Risk* is the probability of harm being caused, usually expressed as a percent or ratio of the treated population; the probability of an occurrence.
- *Harm* is the nature and extent of the actual damage that could be caused. It should not be confused with risk.
- *Effectiveness* is used to express the extent to which a drug works under real world circumstances, i.e., clinical practice (not clinical trials).
- *Efficacy* is used to express the extent to which a drug works under ideal circumstances (i.e., in clinical trials).

20.3 The Specific Aims of Pharmacovigilance

- improve patient care and safety in relation to the use of medicines and all medical and paramedical interventions.
- improve public health and safety in relation to the use of medicines.
- contribute to the assessment of benefit, harm, effectiveness and risk of medicines, encouraging their safe, rational and more effective (including cost-effective) use.
- promote understanding, education and clinical training in pharmacovigilance and its effective communication to the public.

20.4 Definitions

Adverse drug reaction

"A response which is noxious and unintended and which occurs at doses normally used in humans for the prophylaxis, diagnosis or therapy of disease, or for the modification of physiological function." (WHO, 1972)

Karch and Lasagna defined ADR as "Any response to a drug that is noxious and unintended, and that occurs at doses used in humans for prophylaxis, diagnosis, or therapy, excluding failure to accomplish the intended purpose."

Adverse event

"Any untoward medical occurrence that may present during treatment with a pharmaceutical product but which does not necessarily have a causal relationship with this treatment."(WHO)

Seriousness and Severity

The American *Food and Drug Administration* defines a serious adverse event as one when the patient outcome is one of the following

- Death
- Life-Threatening
- Hospitalization (initial or prolonged)
- Disability - significant, persistent, or permanent change, impairment, damage or disruption in the patient's body function/structure, physical activities or quality of life.
- Congenital Anomaly
- Requires Intervention to Prevent Permanent Impairment or Damage

Severity is a point on an arbitrary scale of intensity of the adverse event in question. The terms "severe" and "serious" when applied to adverse events are technically very different. They are easily confused but can not be used interchangeably, require care in usage.

A headache is severe, if it causes intense pain. There are scales like "visual analog scale" that help us assess the severity. On the other hand, a headache can hardly ever be serious, unless it also satisfies the criteria for seriousness listed above.

Side effect: Any unintended effect of a pharmaceutical product occurring at normal dosage which is related to the pharmacological properties of the drug.

Unexpected adverse reaction: An adverse reaction, the nature or severity of which is not consistent with domestic labelling or market authorization, or expected from characteristics of the drug.

Signal: Reported information on a possible causal relationship between an adverse event and a drug, the relationship being unknown or incompletely documented previously. Usually more than a single report is required to generate a signal, depending upon the seriousness of the event and the quality of the information. The publication of a signal usually implies the need for some kind of review or action.

Serious unexpected adverse drug reaction(SUADR):A serious adverse drug reaction that is not identified in nature, severity or frequency by the risk information provided in the clinical investigator's brochure (CIB) or on the drug label.

20.5 ADR Classification and Evolution

Haber (1924) - Haber's Law: the severity of a reaction is a function of concentration and the duration of exposure

Wayne (1958) -Distinguished predictable and unpredictable effects

Lasagna (1964)- Distinguished predictable and unpredictable effects

Levine (1973) - Distinguished dose related and non-dose related reactions and distinguished between acute, subacute, and chronic toxic reactions based on time relationship.

Wade and Beeley (1976)- Distinguished dose related and non-dose related effects and long term and teratogenic effects

Rawlins and Thompson(1977) - Proposed two types of adverse drug reaction:Type A and Type B

Grahame-Smith and Aronson (1984)- Classified types A and B as dose related and non-dose related reactions. Added two time related categories: long term and delayed

Hoigne et al (1990) - Distinguish acute, subacute, and latent allergic reactions

Park et al., (1992) - Labelled Grahame-Smith and Aronson's time related categories as types C (long term) and D (delayed).

Laurence and Bennett (1992) - Split type C into two types, type C (continuous) and type E (end of use).

Ferner and Mann(1997) - Distinguished five patterns of time course.

Aronson and Ferner (2003) - Distinguished three types of dose related adverse reactions: toxic effects, collateral effects, hypersusceptibility and distinguished between time dependent and time independent reactions.

Source: J K Aronson and R E Ferner Joining the DoTS: new approach to classifying adverse drug reactions BMJ, Nov 2003; 327: 1222 – 1225

Rawlins and Thompson, 1977

	Type A(Augmented)	Type B(Bizzare)
Pharmacologically predictable	Yes	No
Dose dependent	Yes	No
Frequency	Common	Rarer
Incidence	High	Low
Mortality	Low	High
Treatment	Adjust Dose	Stop the Drug

Rawlins MD, Thompson JW. Pathogenesis of adverse drug reactions. In: Davies DM, ed. Textbook of adverse drug reactions . Oxford : Oxford University Press, 1977: 10.

Type of reaction	Mnemonics	Features
A: dose related	Augmented	Related to pharmacology (toxic effect or side effect—for example, digoxin toxicity)
B: non-dose related	Bizarre	Unrelated to pharmacology (idiosyncratic for example, malignant hyperthermia, or immunological—for example, penicillin rash) Related to cumulative drug use—for or chronic example, NSAID induced renal failure
C: dose and time related	Continuous or chronic	Apparent only some time after use of drug—for example, thalidomide in first trimester and phocomelia limb defects
D: delayed effect	Delayed	Apparent only some time after use of drug—for example, thalidomide in first trimester and phocomelia limb defects
E: Withdrawal	Delayed	Apparent only some time after use of drug—for example, thalidomide in first trimester and phocomelia limb defects
F: Failure of therapy	Failure	Results from the ineffective treatment (previously excluded from analysis according to WHO definition) e.g: accelerated hypertension because of inefficient control

Edwards IR, Aronson JK. Adverse drug reactions: definitions, diagnosis and management. *Lancet* 2000; 356:1255-9.

Aronson JK. Drug therapy. In: Haslett C, Chilvers ER, Boon NA, Colledge NR, Hunter JAA, eds. *Davidson's principles and practice of medicine* 19th ed. Edinburgh: Elsevier Science, 2002:147-63.

Time related classification of adverse drug reactions

1. Time Independent

Type of reaction	Examples	Implications
Due to change in dose or concentration (pharmaceutical effect)	Toxicity due to increased systemic availability	Beware of changing formulations of some drugs (e.g. modified-release formulations of lithium)
Due to change in dose or concentration (pharmacokinetic effects)	Digitalis toxicity due to renal insufficiency	Forewarn the patient. Monitor carefully throughout treatment. Alter dosage when pharmacokinetics change (e.g. renal insufficiency). Avoid interacting drugs.
Occurs without change in dose (pharmacodynamic effects)	Digitalis toxicity due to renal insufficiency	Forewarn patient. Monitor carefully throughout treatment. Avoid precipitating (pharmacodynamic) factors. Avoid interacting drugs

2. Time dependant

Type of reaction	Examples	Implications
Rapid (due to rapid administration)	Red man syndrome (vancomycin)	Administer slowly
First dose reactions	Hypotension (a 1 adrenoceptor antagonists and angiotensin converting enzyme inhibitors)	Take special precautions for the first dose
Early (abates with repeated exposure)	Adverse reactions that involve tolerance (e.g. nitrate induced headache)	Monitor during early stages. Give appropriate reassurance. Expect adverse effects if strategies to avoid tolerance are adopted.
Intermediate (risk increases at first, then diminishes)	Venous thromboembolism (antipsychotic drugs)	Monitoring not needed after the high risk period unless susceptibility changes. Withdraw drug if a reaction develops.
Late (risk increases with time)	Osteoporosis (corticosteroids)	Assess baseline function. Forewarn the patient. Monitor periodically during prolonged treatment.
Delayed	Carcinogenesis (cyclosporine, diethylstilbestrol)	Avoid or screen. Counsel or forewarn the patient.

Joining the DoTS: new approach to classifying adverse drug reactions. Aronson JK , Ferner RE . BMJ_2003 Nov 22;327(7425):1222-5

Classification based on frequency

Very common	> 1/10	(> 10%)
Common (frequent)	> 1/100 and < 1/10	(> 1% and < 10%)
Uncommon (infrequent)	> 1/1,000 and < 1/100	(> 0.1% and < 1 %)
Rare	> 1/10,000 and < 1,000	(> 0.01% and < 0.1%)
Very rare	< 1/10,000	(< 0.01%)

Based on avoid ability of the ADR s

Definitely avoidable	The ADR was due to a drug treatment procedure inconsistent with present day knowledge of good medical practice
Possibly avoidable	The ADR could have been avoided by an effort exceeding the obligatory demands of present day knowledge of good medical practice
Unavoidable	The ADR could not have been avoided by any reasonable means

Hallas J, Harvald B, Gram LF, Grodum E, Prosen K, Haghfelt T, et al. Drug related hospital admissions: the role of definitions and intensity of data collection, and the possibility of prevention. *J Intern Med* 1990; 228: 83-90

Based on severity

Severity	Description
Mild	No antidote or treatment is required; hospitalization is not prolonged
Moderate	A change in treatment (e.g., modified dosage, addition of a drug), but not necessarily discontinuation of the drug, is required; hospitalization may be prolonged, or specific treatment may be required
Severe	An ADR is potentially life threatening and requires discontinuation of the drug and specific treatment of the ADR
Lethal	An ADR directly or indirectly contributes to a patient's death

Grading adverse events

By this grading scale, all adverse events are classified as follows:

0 = No adverse event or within normal limits

1 = Mild adverse event

2 = Moderate adverse event

3 = Severe and undesirable adverse event

4 = Life-threatening or disabling adverse event

5 = Death related to adverse event

20.6 Causality Assessment Scales

Causality assessment is the method by which the extent of relationship between a drug and a suspected reaction is established

Currently wide variety of causality assessment scales exist, to attribute clinical events to drugs in individual patients or in case reports, each with their own advantages and limitations. These scales include

Karch & Lasagna scale

Naranjo's scale

WHO probability scale

Spanish quantitative imputation scale

Kramer's scale

Jones scale

European ABO system

Bayesian system.

The Naranjo's scale and the WHO scale of assessment are the most commonly used scales.

Naranjo's Algorithm

Question	Yes	No	Don't know
Are there previous conclusion reports on this reaction?	+ 1	0	0
Did the adverse event appear after the suspect drug was administered?	+ 2	− 1	0
Did the AR improve when the drug was discontinued or a specific antagonist was administered?	+ 1	0	0
Did the AR reappear when drug was readministered?	+ 2	− 1	0
Are there alternate causes [other than the drug] that could solely have caused the reaction?	− 1	+ 2	0
Did the reaction reappear when a placebo was given?	−1	+1	0
Was the drug detected in the blood [or other fluids] in a concentration known to be toxic?	+ 1	0	0
Was the reaction more severe when the dose was increased, or less severe when the dose was decreased?	+ 1	0	0
Did the patient have a similar reaction to the same or similar drugs in any previous exposure?	+ 1	0	0
Was the adverse event confirmed by objective evidence?	+ 1	0	0

Source: A METHOD FOR ESTIMATING THE PROBABILITY OF ADVERSE DRUG REACTIONS - **NARANJO et al**

SCORING FOR NARANJO's ALGORITHM

- > 9 = definite ADR
- 5-8 = probable ADR
- 1-4 = possible ADR
- 0 = doubtful ADR

Who Scale of Assessment

1.Certain	A clinical event, including laboratory test abnormality, occurring in a plausible time relationship to drug administration, and which cannot be explained by concurrent disease or other drugs or chemicals. The response to withdrawal of the drug (dechallenge) should be clinically plausible. The event must be definitive pharmacologically or phenomenologically, using a satisfactory rechallenge procedure if necessary.
2.Probable/ Likely	A clinical event, including laboratory test abnormality, with a reasonable time sequence to administration of the drug, unlikely to be attributed to concurrent disease or other drugs or chemicals, and which follows a clinically reasonable response on withdrawal (dechallenge). Rechallenge information is not required to fulfil this definition.ical event, including laboratory test abnormality, with a reasonable time sequence to administration of the drug, unlikely to be attributed to concurrent disease or other drugs or chemicals, and which follows a clinically reasonable response on withdrawal (dechallenge). Rechallenge information is not required to fulfil this definition.
3.Possible	A clinical event, including laboratory test abnormality, with a reasonable time sequence to administration of the drug, but which could also be explained by concurrent disease or other drugs or chemicals. Information on drug withdrawal may be lacking or unclear.
4.Unlikely	A clinical event, including laboratory test abnormality, with a temporal relationship to drug administration which makes a causal relationship improbable, and in which other drugs, chemicals or underlying disease provide plausible explanations.

Source : WHO

20.7 Important of Post Marketing Surveillance and ADR Reporting

Tests in animals are insufficient to predict human safety, Patients used in clinical trials are selected and limited in number, the conditions of use differ from those in clinical practice and the duration of trials is limited, By the time of licensing exposure of less than 5000 human subjects to a drug allows only the more common ADR to be detected, At least 30,000 people need to be treated with a drug to be sure that you do not miss at least one patient with an ADR which has an incidence of 1 in 10,000 exposed individuals, Information about rare but serious adverse reactions, chronic toxicity, use in special groups (such as children, the elderly or pregnant women) or drug interactions is often incomplete or not available, Therefore reporting of an adverse event by healthcare professionals can save lives of many patients. Source (WHO, 2002)

20.8 Characteristics of a Good Adverse Drug Reaction Report

Good case reports include the following elements:

(i) Description of the adverse events or disease experience, including time to onset of signs or symptoms.

(ii) Suspected and concomitant product therapy details (i.e., dose, lot number, schedule, dates, duration), including over-the-counter edications, dietary supplements, and recently discontinued medications.

(iii) Patient characteristics, including demographic information (e.g., age, race, sex), baseline medical condition prior to product therapy, co-morbid conditions, use of concomitant medications, relevant family history of disease, and presence of other risk factors.

(iv) Documentation of the diagnosis of the events, including methods used to make the diagnosis.

(v) Clinical course of the event and patient outcomes (e.g., hospitalization or death).

(vi) Relevant therapeutic measures and laboratory data at baseline, during therapy, and subsequent to therapy, including blood levels, as appropriate.

(vii) Information about response to dechallenge and rechallenge.

(viii) Any other relevant information (e.g., other details relating to the event or information on benefits received by the patient, if important to the assessment of the event).

SOURCE (FDA)

20.9 Periodic Safety Update Report (PSUR)

A PSUR is intended to provide an update of the worldwide safety experience of a medicinal product to Competent Authorities at defined time points post-authorisation. At these times, marketting authorisation holders are expected to provide succinct summary information together with a critical evaluation of the risk-benefit balance of the product in the light of new or changing information. This evaluation should ascertain whether further investigations need to be carried out and whether changes should be made to the marketing authorisation and product information.

PSURs must be submitted for all registered products regardless of marketing status. A single report may cover all products containing the same active substance(s) licensed by one MA holder.

The report will usually include all dosage forms and formulations, as well as all indications, associated with such an active. Within the PSUR, separate presentations of data for different dosage forms, indications or populations (for example, children vs. adults) may be appropriate, however an overview of the combined data should also be provided.

For combinations of substances which are also registered individually, safety information for the fixed combination may be reported either in a separate PSUR or be included as a separate presentation in the report for one of the separate components, depending on the circumstances. Cross-referencing all relevant PSURs is essential

Frequency of PSUR reporting requirements PSURs should be submitted at the following times from the time of authorisation, for all medicinal products unless the marketing authorisation makes different provisions:

- Immediately upon request
- At least 6 monthly after authorisation and until the placing on the market
- At least 6 monthly for the first two years after being placed on the market
- Annually for the subsequent two years
- Thereafter at three-yearly intervals

Multiples of six-monthly PSURs are acceptable, provided that the MA holder submits a PSUR bridging summary report (BSR). Where multiple reports are being provided the BSR should be comprehensive

Sources of information Generally data from the following sources of adverse reaction case information are potentially available to a marketing authorisation holder and should be included in the PSUR:

1. Direct reports to authorisation holder :
   - spontaneous notifications from health-care professionals
   - MA holder sponsored clinical studies or named patient (compassionate) use
   - spontaneous notifications from consumers/patient
2. Literature
3. ADR reporting systems of regulatory authorities:
   - spontaneous notifications and non-spontaneous notification
4. Other sources of data:
   - reports of adverse reactions exchanged between contractual partners (e.g. licensors-licensees)

- data in special registries, such as maintained in organ toxicity monitoring centres
- reports created by poison control centres
- epidemiological databases

20.10 MedDRA or Medical Dictionary for Regulatory Activities

MedDRA or Medical Dictionary for Regulatory Activities is a clinically validated international medical terminology used by regulatory authorities and the regulated biopharmaceutical industry throughout the entire regulatory process, from pre-marketing to post-marketing activities, and for data entry, retrieval, evaluation, and presentation. In addition, it is the adverse event classification dictionary endorsed by the International Conference on Harmonisation of Technical Requirements for Registration of Pharmaceuticals for Human Use (ICH). MedDRA is used in the US, European Union, and Japan. Its use is currently mandated in Europe and Japan for safety reporting. Thus MedDRA was developed to overcome some of the limitations of previous AE/ADR coding terminologies and to provide a single, multilingual resource for the classification of AE/ADR data. Currently, MedDRA's use is mandated for various types of regulatory reporting in the EU and Japan; it is widely used by the biopharmaceutical industry in all three ICH regions (US, EU and Japan). The US Food and Drug Administration has used MedDRA in its Adverse Event Reporting System for over a decade.

MedDRA provides a robust, vigorously maintained standard terminology for AE/ADR reporting with sufficient granularity of concepts and logical medical groupings within its hierarchy of terms. However, because of the size and specificity of MedDRA, as well as the organization of some of its groupings, retrieval and aggregation of coded AE/ADR data can sometimes present a challenge.

MedDRA is managed by the MSSO (Maintenance and Support Services Organization), an organization that reports to the International Federation of Pharmaceutical Manufacturers and Associations (IFPMA). MedDRA is free for regulators and priced according to company revenue for industry. MedDRA is also available in Japanese. The Japanese counterpart for MSSO is called JMO.

The MSSO updates MedDRA from subscriber change requests to add a new medical concept that is yet to be in MedDRA or to change an existing concept. The decisions are made by international medical officers on how to map the terminology within the grouping categories according to a general consensus based on language considerations internationally

20.11 Pharmacovigilance in India

The National Pharmacovigilance Programme at CDSCO aims to:

1. monitor the adverse drug reactions of medicines in order to identify previously unexpected adverse drug reactions or indicate that certain reactions occur more commonly than previously believed. This will include the collation, review and evaluation of all spontaneous ADR reports received by the unit on a nation-wide basis. This information will then be keyed into the ADR database for use in aggregate analysis. These reports shall also be submitted to the WHO International Drug Monitoring Programme for international collaboration on drug safety.
2. review Periodic Safety Update Reports (PSURs) submitted by pharmaceutical companies. Pharmaceutical companies are required to submit the PSURs of all new chemicals drugs. PSURs shall be expected to be submitted every 6 monthly for the first 2 years of marketing in India, and annually for the subsequent 2 years.
3. maintain contacts with international regulatory bodies working in pharmacovigilance and exchange information on drug safety.
4. assess the regulatory information relating to safety in order to determine what action, if necessary, needs to be taken to improve safe use. Based on the available data, the Advisory Committee shall make recommendations on product label amendments, product withdrawals and suspension.
5. provide information to end-users through adverse drug reaction news bulletins, drug alerts and seminars.

SOURCE: CDSCO *(http://cdsco.nic.in/)*

Importance of ADR reporting in India

Adverse Drug Reactions are fourth to sixth leading cause of death among hospitalised patients and it occurs in 0.3 per cent to 7 per cent of all hospital admissions. The incidence of serious ADRs is 6.7 per cent (Joshi et al).

There is a rapid increase in the number of new drugs entering the market from last few decades, India being the second most populated country has over one billion potential drug consumers, and no amont of pre-clinical and clinical data is sufficient to conclude the complete safety of a drug, under this scenario it becomes necessary to report any untoward reaction of any pharmaceutical product to assess its safety and efficacy to ensure maximal patient health.

In India, general practitioners, with an large outpatient base tend to be among the first ones to use the drugs entering the market, hence they are in the best position to assess the adverse drug reactions associated with drugs.

Suggested Reading

1. **CDSCO** *(http://cdsco.nic.in/)*
2. Hallas J, Harvald B, Gram LF, Grodum E, Prosen K, Haghfelt T, et al. Drug related hospital admissions: the role of definitions and intensity of data collection, and the possibility of prevention. *J Intern Med* 1990;228: 83-90
3. J K Aronson and R E Ferner. Joining the DoTS: new approach to classifying adverse drug reactions BMJ, Nov 2003; 327: 1222 – 1225.
4. Rawlins MD, Thompson JW. Pathogenesis of adverse drug reactions. In: Davies DM, ed. Textbook of adverse drug reactions. Oxford : Oxford University Press, 1977: 10.
5. Edwards IR, Aronson JK. Adverse drug reactions: definitions, diagnosis and management. *Lancet* 2000; 356:1255-9.
6. Aronson JK. Drug therapy. In: Haslett C, Chilvers ER, Boon NA, Colledge NR, Hunter JAA, eds. *Davidson's principles and practice of medicine* 19th ed. Edinburgh: Elsevier Science, 2002:147-63.

CHAPTER 21

Data and Safety Monitoring Board and Monitoring Plan

P. Usha Rani and M.U.R. Naidu

21.1 Definition and Purpose

Data and Safety Monitoring Board (DSMB) is called Data Monitoring Committee (DMC). Clinical research with investigational products, devices, procedures etc involving human subjects is designed to answer specific questions about human physiology or effects of a particular intervention on human body. Data and safety monitoring is the process for reviewing accumulated outcome data from an ongoing clinical trial to ensure the continuing safety of current participants and those yet to be enrolled, as well as the continuing validity and scientific standards of the research project. Thus, DMC is an independent group of experts who monitor patient safety and treatment efficacy data in an on-going clinical trial.

An independent DSMB is the most appropriate way to monitor data and safety for studies that involve the following:

(a) Large number of participants
(b) Blinded study treatment groups
(c) Multiple clinical study sites
(d) High risk interventions

(e) Controlled trials with mortality or major morbidity as a primary or secondary endpoint.

21.2 Data Monitoring Committee

The DSMB is an independent committee, external to the trial organizers and investigators, formally established to monitor data throughout the planned study, appropriately, on scientific and ethical aspects.

Members of DSMB- Typically, DMC consists of 3-7 members (preferably odd numbers and persons not directly involved with the design and conduct of study) which include a statistician, clinicians knowledgeable about the disease, also, clinicians who have knowledge in fields of major safety effects like cardiology, neurology, nephrology etc. Long term trials may include an ethicist or a representative from patient advocacy group as a member.

Qualifications of members – Membership includes – 1. Expert in the field 2. Experience in conduct of clinical trials and statistical knowledge 3. independent from the direct management of the clinical study and 4. no conflict of interest. A chairperson appointed will be the contact person of DMC and is responsible for overseeing the meetings, develop the agenda and summarize the meeting.

Responsibilities of DMC – The DSMB should meet to review the research protocol, informed consent documents, safety and data monitoring plan prior to enrolment of first subject. Te board also decides on the data that will be required for review. They further decide the schedule of meetings, appoint the chair, who receives the minutes and signoff conflict of interest statements.

The DSMB evaluates progress of the study, including recruitment, protocol adherence, retention of subjects and also periodic assessment of data for quality and completeness. They may monitor the scientific integrity of study, timelines of accrual, the quality of data collection and management and the outcomes to assure safety of study subjects. Further, interim analysis of data is conducted by DMC for efficacy or safety assessment to determine whether the trial should continue as originally planned, changed or stopped based on the data. Typically DMC may recommend termination of the study based on 1. safety concerns 2. outstanding benefit and 3. futility.

Last but not the least, DSMB protects confidentiality of the study participants, data and the monitoring results.

Timing and frequency of meeting

The DMC will meet at least annually or more often at predetermined intervals (3-6 months) and review data. The committee may meet periodically if the risk to the subject is high, the population is vulnerable, there is a large volume of data to review, and/or after a specified number of subjects are enrolled in the study. Further, the Chairman may call for ad-hoc meetings and/or teleconferences depending on safety and efficacy issues.

21.3 Data Safety Monitoring Plan

The data and safety monitoring plan (DSMP) is normally formulated based on nature, size and complexity of the research study, the expected risks and type of population under study. The DSMP normally includes the following:

1. Type of data or events captured under monitoring plan.
2. Who is responsible for monitoring the data collected, including data related to unexpected adverse events.
3. Evaluate aggregate analysis of adverse events and serious adverse events.
4. Time frames for reporting adverse events to monitoring team
5. Adherence to treatment plan
6. Review interim analysis where applicable, determine specific data to be analyzed.
7. Frequency of assessment of data captured by monitoring plan.
8. Defining stopping rules, which determine that study should be altered or stopped. Stopping rules should be specific about the endpoints that will be used and the decisions that will be made.
9. Review protocol violations and deviations to assess adequacy of study.
10. Ensure documentation of informed consent process
11. Review subject enrolment – followed eligibility criteria; number of subjects enrolled; compliance to visits; information on screen failures.
12. Review IND information
13. Review completeness and quality of data
14. Review vital signs; clinical and laboratory tests.
15. Review confidentiality
16. As appropriate, procedures for communicating to the IRB, sponsor and other appropriate personnel the outcome of reviews by monitoring team.

Determination of study risk and monitoring plan based on study risk:

Risk assessment should include consideration of both the intervention being performed and study population and also special circumstances that are unique to the study like genetic testing etc.

Minimal risk – A minimal risk study is one where the probability and magnitude of harm / discomfort in the proposed study is not more than that encountered in daily life or performance of routine procedures like venipuncture, exercise testing, DEXA scans, oral glucose tolerance tests etc. Minimal risk does not require an independent DSMB and the monitoring in such situations is the responsibility of the investigator.

Moderate risk – Moderate risk studies include Phase I-III studies using drugs for their indicated use, studies requiring invasive hemodynamic monitoring with arterial lines, storing of tissue and blood samples, genetic studies which require blood to be withdrawn for real time or stored for later use. In most cases independent safety monitoring is required for studies with moderate risk, especially if he study is placebo-controlled or an investigational agent is used.

High risk – High risk studies include clinical trials with investigational agents, gene transfer studies, new indication of approved drugs, Phase I clinical trials, investigator initiated IND's (investigational new drugs) or IDE's (investigational devices); some Phase II clinical trials also included in this category are procedures like interventional radiology, studies in which subjects may be at risk of death or severe morbidity because of existing illness or disease. High risk require DSMB to be constituted. In some high risk situations, monitoring may be performed by an independent investigator who is an expert in the agent being studied and the patient population.

Conclusion

Regardless of data and safety monitoring plans by the sponsor or others, the ultimate responsibility for protecting rights, safety and welfare of the subjects rests on the principal investigator. It is the principal investigator's responsibility to ensure that the study is conducted at their site in accordance with IRB approval and applicable regulatory requirements.

Appendix

Roles and Responsibilities

DSMB Roles and Responsibilities

This DSMB will

- Meet periodically meet to review aggregate and individual subject data related to safety, data integrity and overall conduct of the trial.
- Review specific interim analyses for efficacy
- Provide recommendations to continue or terminate the trial depending upon these analyses.
- Communicate other recommendations or concerns as appropriate.
- Operate according to the procedures
- Follow conflict of interest guidelines
- Comply with confidentiality procedures
- Maintain documentation and records of all activities as described below

Sponsor (or Designees) Roles and Responsibilities

- The sponsor will directly or through delegation:

- Assure the proper conduct of the study.
- Assure collection of accurate and timely data (monitoring and data management).
- Compile and report SAEs to the DSMB.
- Promptly report potential safety concern(s) to the DSMB.
- Prepare summary reports of relevant data for the DSMB.
- Provide an independent facilitator for presentation of results during DSMB meetings if requested by the DSMB.
- Communicate with regulatory authorities, IRB/EC, and investigators, in a manner that maintains integrity (e.g., blinding) of the data, as necessary. (This communication is not the responsibility of the DSMB.)
- Provide funding for the study and DSMB.

Suggested Reading

1. Bert Spilker. Guide to clinical trials. New York. Raven Press 1999.
2. Donald Cutlip, MD, James Tcheng, MD, Rebecca Budreau, Neal Fearnot, Scott Snyder, Willis Tacker. Chartering a Data Safety Monitoring Board. Applied Clinical Trials. Jul 1, 2008.
3. Food and Drug Administration, *Guidance for Clinical Trial Sponsors on the Establishment and Operation of Clinical Trial Data Monitoring Committees* (FDA, Rockville, MD, March 2006).
4. S. Ellenberg, T. Fleming, D. DeMets, "Responsibility of the Data Monitoring Committee and Motivating Illustrations," in *Data Monitoring Committees in Clinical Trials: A Practical Perspective*, S. Ellenberg, T. Fleming, and D. DeMets (John Wiley and Sons Ltd., West Sussex, England, 2002), pp. 19-43.
5. DAMOCLES Study Group, "A Proposed Charter for Clinical Trial Data Monitoring Committees: Helping Them to Do Their Job Well," *Lancet*, 365, 711-722 (2005).

CHAPTER 22

Standard Operating Procedures

U. Shobha Jagadish Chandra

:2.1 Introduction

:linical research has grown over the last fifty years. Biomedical research requires pecialized knowledge, which includes regulatory requirements, good clinical ractice (GCP) and medical ethics. Failure to understand the intricacies of onducting clinical trials may lead to serious errors in implementation of a study nd compromise the ensuing data. The interpretation of the intervention effects nay be flowed and therefore, unacceptable to regulatory authorities.

Food and drug administration (FDA) regulations define the fundamentals of ;CP, there are numerous other guidelines and codes that also dictate standards 1 the conduct of clinical research. For example, the Nuremberg Code serves as 1e basis for the protection of human subjects. When combined, these concepts erve as the gold standard in the practice of clinical research.

In recent years, international GCP standards have emerged in an effort to armonize research efforts, reduce waste and expedite approval. The United tates, European Union and Japan have joined forces at the International :onference on Harmonization (ICH) to develop global standards for the conduct f clinical research. The purpose of ICH is to facilitate mutual acceptance of ata submitted in support of drug marketing applications. Their GCP guideline as been adopted by industry and academia worldwide. With all countries conducting research according to the same standards, duplication of effort and resources will be minimized.

GCP is defined as "A standard for the design, conduct, performance, monitoring, auditing, recording, analyzing and reporting of clinical trials that provides assurance that the data and reported results are credible and accurate and that the rights, integrity and confidentiality of trial subjects are protected.

FDA and other Federal regulatory authorities have halted research activities at some of the major academic institutions due to non-compliance with regulations involving data integrity, protection of human subjects and conflicts of interest. The clinical research enterprise will never escape from regulation. It is the responsibility of the Sponsor, the Investigator and their institutions to establish an institutional culture where studies are conducted in an ethical and scientifically rigorous manner and where safeguards to subjects who participate in clinical studies are of paramount importance. The first step in establishing an institutional culture of research excellence is through the establishment of operational policies and standard operating procedures that address the regulatory requirements, ethical conduct and sound implementation of all research conducted on human subjects.

22.2 Standard Operating Procedures

The organization of SOP is intended to group together related activities and / or concepts.

Standard Operating Procedure (SOP) is defined as "standard, elaborate, written instructions to achieve uniformity of performance in the management of clinical studies." SOPs provide a general framework for the efficient implementation and performance of all the functions and activities related to a particular study. SOPs provide a general framework for the efficient implementation and performance of all the functions and activities related to a particular study.

SOPs are standard set of instructions

- Defining purpose, scope and how a task is to be performed to achieve a specified outcome.
- Standardizing the way a task is to be performed to minimize variation.
- Forming the basis for training the "task" performers.
- Providing an informational tool that supports the performers.
- Fulfilling regulatory requirements.

Sound designing of the SOP's leads to success of the project. SOPs are an essential element of quality assurance and is therefore imperative that they are understood and followed by all end users in a clinical trial. If SOPs are not followed correctly, the validity of data generated can be compromised leading to delays in the drug development process. From regulatory point of view, SOPs, provide a written record of the process followed in the clinical trial, ensure compliance with regulations and guidelines, ensure quality of data, facilitate audit and inspections and assure global acceptance, for auditors and inspectors. Therefore, SOPs become mandatory to set benchmarks and standards.

From the point of view of the task to be accomplished, SOPs ensure consistency of task implementation and therefore enhanced performance while for managers SOPs are a useful training tool, helping in generating confidence in team members, while at the same time they reduce supervisory time / effort. For quality assurance personnel, SOPs are crucial for performing audits.

SOPs are useful when the activity / task is important and more than one person is involved in the task, the performance of the task can affect the service / product and task needs to be done repeatedly and consistently.

The most appropriate person to write an SOP would be an in-house person who has designed the procedure and its controls and who actually performs the task routinely or who is directly responsible for the performance of that task. Knowledge and familiarity about the process and good communication skills are additional qualification required to write an SOP.

SOPs have to be comprehensive, practical and customized. SOP should be easy to understand, execute / implement and also allow changes in the future. SOPs should be written so that, they are dynamic and flexible to future changes yet robust enough to allow easy transfer to other personnel or laboratories. Each organization and departments in that organization needs to generate its own SOPs depending on the functions or activities it performs.

SOPs are more than simple functional documents that define a task or tell what to do or how to do it. Although not fundamentally different in content from other procedural documents like instruction manual. They are distinctive in format and exist in co-coordinated sets that demand rigorous management. The writer of SOPs plays a critical role in an organization, because he or she is defining the activities of the organization itself.

Steps to write SOPs.

1. Generation (identify the steps and create a flow chart, write the draft SOP)
2. Review and approval
3. Distribution
4. Training
5. Implementation
6. Maintenance

Generation of SOP

It is useful to first study and understand the procedures for which SOP is to be written, examine whether any previously written SOPs are available in the organization, be clear about the experience and qualifications of the users and how the SOP will be used.

First the writer should know the task for which the SOP is being written. The objective, scope and responsibilities, the flow chart of the steps that form the entire task, and the detailed steps that will lead to the successful completion of the task.

A simple flow chart describing what is done at each step including the critical "hows" must be written. From that sketch, it becomes easy to write the procedure using the "Official" template. All written SOPs contain at the bare minimum the following and each has to be pre-decided in the planning of the SOP.

Category	:	(Administrative / Study Conduct / Pre-Study / Ethical Review)
Title	:	Which should be descriptive
SOP number	:	Each organization will develop its own system of numbering the SOPs
Version	:	??
Total Pages	:	(Each page has to be numbered as page – of -)
Validity	:	(First effective date and review date)
Author	:	
Reviewer	:	

Approved by with date of approval

Review and approval of an SOP.

After the draft SOP is written, it is useful to get it read by others who may be performing or supervising the task or talking through the steps in a small group of three or four people. The objective is to ensure that the draft document is complete, correct and "do-able". After this the SOP is formally reviewed. There are two types of review : **inspection review** which basically confirms that the SOP conforms to pre-decided standards - does it have a unique title and number, are all pages and attachments present etc., and **content review** which requires subject expertise.

Distribution

Distribution to who is going to need to know about the SOP apart from those who will be responsible for execution of the task. For example, if the SOP relates to acceptance of an Ethics Committee application, although the EC Secretariat is responsible for the task, the SOP must be made available to all potential investigators who will submit the application to facilitate implementation. Or if the SOP is for administering and documenting an informed consent, it will apply to the PI study co-ordinator but may be distributed to the entire study team.

Training

Training to the individuals who will be using and managing SOPs is by far the most important step in the management of SOP. This has to be completed before the effective date of procedure. The training process must be documented.

Implementation

The SOP is implemented on the date specified after distribution and training of the concerned staff.

Maintenance

Generally, an SOP is a controlled document, in the sense that it requires signature approval. Revisions are numbered and approved as well. There should be a mechanism to ensure that only the latest approved version is in circulation. The earlier versions must be retrieved and archived. If an organization for whom the SOP is written for, does not have a document management system, it would mandate that there be someone responsible for these to control. Archival of SOPs is also important.

22.3 SOP Template

Index

Objective: This states the aim, content and general description of the SOP.

Responsibilities: This describes who are the personnel responsible for execution of the tasks described in the SOP (it is useful to involve the people who are responsible for the task and who will actually use the SOP while writing the pertinent SOP as it gives them all a sense of ownership and participation, making training and implementation of SOPs easy).

Scope: This sets the limits to the use of the SOPs stating to what areas this procedure does and does not apply.

Reference to other applicable SOPs: If there are any other SOPs which may provide additional information on the subjects described in the current SOP, then it is described here.

Appendices: if there are any important attachments which are needed for execution of the SOP then it is described here. It is important to label them sequentially.

Flow chart: This summarizes the steps that will be described in detail in the next section. This allows for at a glance understanding of the tasks described in the SOP.

Procedure or instructions: A step-by-step description with instructions for completing the task is given here.

Steps to Good Documentation Practice

1. Information recorded should be easy to read (i.e Legible)
2. Abbreviated date should be recorded by Day / Month / Year
3. Blue or Black ball pen / ink pen must be used when completing documents.
4. All defined spaces which are not used must be completed with "N/A" dated and signed by responsible colleague.
5. Scrap paper should not be used for recording the data.
6. Any document must never be post dated or back dated.
7. If any error is found immediately after recording the data, cross out incorrect information with a single line. Enter correct data, sign and date near correction.
8. Information should be recorded on proper forms, laboratory note books, batch manufacturing records and cleaning records, immediately after actions are completed. Entries of signature or initials and dates must be made at the same time.
9. Do not overwrite any data and blank space in-process sheets, batch manufacturing records log books and laboratory note books.
10. Information should be recorded on the documents concurrent to the activity.
11. Corrective fluid or erasable pens should not be used to correct the original information.
12. Original information should always be entered directly to the applicable document.

22.4 Master Formulation Record

Objective

To assure uniformity master formula (master production and control) records for each product, including size shall be prepared.

Procedure

It is to be described in a written procedure and such written procedure shall be dated and signed by competent technical staff and independently checked, dated and signed by second person. Master formula record shall include.

1. Name and strength of the product along with dosage form (if it is a pharmacopoeial product under trade name, both names will be given)
2. The name and weight / measure of active ingredients per dosage unit or per unit weight / measure of the product and total weight / measure of any dosage unit.
3. A complete list of all the ingredients to be used in the manufacture of the product indicating any special quality characteristics.
4. An accurate statement of weight / measure of each ingredient (same weight system should be used) required as per formula of dosage form and the weight / measure actually to be used (e.g. overages to compensate losses during storage)
5. A statement of theoretical weight / measure at appropriate phases of processing.
6. A statement of theoretical yield including permissible limits beyond which investigation is required.
7. A description of containers, closures and packing materials to be used.
8. A description of all vessels, equipment to be used.
9. Processing and packaging procedures.
10. In process control to be exercised during processing and packaging.
11. Precaution to be taken during manufacture and storage of semi-finished product.
12. Complete manufacturing and control instructions, sampling and testing procedures, specifications, special notations and precautions to be followed.

	STANDARD OPERATING PROCEDURE	**Document Number** CSOP –XX-X	**Rev Level**
(Insert company LOGO) (Insert Company Address)	**Confidential** THIS DOCUMENT AND ALL INFORMATION CONTAINED HEREIN IS PROPRIETARY TO COMPANY AND IS NOT TO BE DISCLOSED WITHOUT THE EXPRESS WRITTEN PERMISSION OF COMPANY	**PAGE** 1	**Total Pages** 7
Description MANAGEMENT OF CLINICAL STUDY DOCUMENTS			

Title: Management of Clinical Study Documents

CSOP-XX-X

Revision History

Revision	Active Date	Initials	Detailed Description of Change

Functional Area	Approved By	Date
INITIATOR		

(Insert company LOGO) (Insert Company Address)	**STANDARD OPERATING PROCEDURE**	**Document Number** CSOP –XX-X	**Rev Level**
	Confidential THIS DOCUMENT AND ALL INFORMATION CONTAINED HEREIN IS PROPRIETARY TO COMPANY AND IS NOT TO BE DISCLOSED WITHOUT THE EXPRESS WRITTEN PERMISSION OF COMPANY	**PAGE** 2	**Total Pages** 7
Description MANAGEMENT OF CLINICAL STUDY DOCUMENTS			

1. Purpose

This procedure describes the regulatory requirements and procedures for management of investigational study documents, including record retention, at the investigation site.

2. Scope

This procedure applies to all Investigational Device Exemption, Post-market Surveillance and other FDA mandated clinical studies conducted by the Company.

3. Terminology Definitions

N/A

4. Responsibilities

The Clinical Manager is responsible for ensuring the suitability and implementation of this procedure. Company personnel who are involved in setting up and / or maintaining study records are responsible for following company Procedure. The Clinical Manager in conjunction with the Regulatory Department is responsible for ensuring that investigational study documents are generated, reviewed and maintained. The Clinical Manager is responsible for ensuring that participating investigators and other study site personnel are aware of and agree to comply with all regulations governing the maintenance and retention of study documents.

5. Applicable Documents

Record Retention Responsibility table (Appendix I)

	STANDARD OPERATING PROCEDURE	Document No CSOP –XX-X	Rev Level
(Insert company LOGO) (Insert Company Address)	Confidential. *THIS DOCUMENT AND ALL INFORMATION CONTAINED HEREIN IS PROPRIETARY TO COMPANY AND IS NOT TO BE DISCLOSED WITHOUT THE EXPRESS WRITTEN PERMISSION OF COMPANY*	Page 3	Total Pages 7
Description MANAGEMENT OF CLINICAL STUDY DOCUMENTS			

6. Procedure

6.1 Investigati onal Study Documents Company

All investigational studies will have two types of files: central study files and individual site study files. The central study file will contain information relevant to all sites in a multi-center trial. The individual site study files will contain information relevant only to that site and will exist for each site participating in a clinical trial.

6.1.1 The central study files will be maintained throughout the study and will include, but not be limited to, the following:

- Approved original investigational plan;
- Technical manuals
- Protocol amendment(s)
- Any general study-related memos / documents (.eg., numbered memos, news letters)
- Copy of the final report when it becomes available;
- Internal correspondence;
- Core laboratory signed investigator agreement (if applicable);
- Core laboratory reference ranges (if applicable)

6.1.2 Individual site study files will be maintained for each site participating in a particular clinical trial. The files will be maintained throughout the study and will include, but not be limited to the following:

- Signed original investigational plan;
- Signed investigational study agreements (including financial disclosure),
- Indemnification agreements (if applicable)
- Curriculum vitae for the principal investigator, all sub-investigators, study coordinators (s)
- Signature log;
- IRB approval letter(s)
- IRB approved informed consent form;
- IRB membership roster(s) & Laboratory certification (if applicable)
- Laboratory reference ranges (if applicable)
- Correspondence to and from site (including correspondence regarding financial issues)

	STANDARD OPERATING PROCEDURE	Document Number CSOP –XX-X	Rev Level
(Insert company LOGO) (Insert Company Address)	**Confidential** THIS DOCUMENT AND ALL INFORMATION CONTAINED HEREIN IS PROPRIETARY TO COMPANY AND IS NOT TO BE DISCLOSED WITHOUT THE EXPRESS WRITTEN PERMISSION OF COMPANY	**Page** 4	**Total Pages** 7
Description MANAGEMENT OF CLINICAL STUDY DOCUMENTS			

- Monitoring reports
- Subject log
- Site visit log
- Training documents
- Device accountability log
- Randomization list / codes (if applicable)
- Copy of the final study report submitted by the principal investigator to the IRB and sponsor.
- Copies of appropriate company check requests and individual checks.
- Contracts (investigative site, CRO, core laboratory)

6.2 Investigational study documents – Site

All the clinical study sites will maintain investigational study files(also called administrative files or regulatory binder). The files will be maintained throughout the study and will contain all or any of the following depending on the study, the sponsor's requirements, and the requirements of the site's IRB.

- Subject log; Signed protocol and all amendments
- Technical manuals; Laboratory values and certification (if applicable)
- Investigator, sub-investigators, and study co-coordinator curriculum vitae (s)
- Signed investigator agreement; IRB approvals; IRB progress reports; IRB renewals; Approved informed consent documents; IRB membership roster
- Device accountability log; Site visit log; Correspondence to and from Company USA; Correspondence to and from IRB

6.3 Record Retention and Archiving

All records pertaining to an investigational study must be retained for two years after the approval by the FDA or if a study does not receive FDA approval, two years after the study has been terminated and the FDA has been so notified.

6.4 Availability of study records for FDA Inspections

Copies of all records in these files must be stored at (company) and at investigational site, or if on site storage is impractical, in a secure off-site facility where they may be readily accessible in the event of an inspection by FDA or other regulatory body where they may be readily accessible in the event of an inspection by FDA or other regulatory body.

	STANDARD OPERATING PROCEDURE	Document Number CSOP –XX-X	Rev Level
(Insert company LOGO) (Insert Company Address)	**Confidential** THIS DOCUMENT AND ALL INFORMATION CONTAINED HEREIN IS PROPRIETARY TO COMPANY AND IS NOT TO BE DISCLOSED WITHOUT THE EXPRESS WRITTEN PERMISSION OF COMPANY	**Page** **6**	**Total Pages** **7**
Description MANAGEMENT OF CLINICAL STUDY DOCUMENTS			

APPENDIX I

Record Retention Responsibility

Significant Risk Device

Record	Maintained by	
	Investigator	Sponsor
All correspondence pertaining to the investigation	X	X
Shipment, receipt, disposition	X	X
Device administration and use	X	
Subject case histories	X	
Informed consent	X	
Protocols and reason for deviations from protocol	X	X
Adverse device effects and complaints	X	X
Signed investigator Agreements	X	X
IRB membership lists/employment/conflicts of interest.		*
Minutes of IRB meetings	*	*

- if possible

Non-significant Risk Device

Record	Maintained by	
	Investigator	Sponsor
Name and intended use of device		X
Brief explanation of why device does not involve significant risk		X
Name and address of investigator(s) and IRBs		X
Degree of GMPs followed		X
Informed Consent	X	
Adverse device effects and complaints		X

Standard operating procedures

The Clinical research Unit (CRU) is accountable for following the regulations set forth by the Food & Drug Administration (FDA) in the investigation of test articles in humans. Per the International Conference on Harmonization: Good Clinical Practice (ICH Guidelines for GCP), CRU need to have Standard Operating Procedures (SOPs) in order to "achieve uniformity of the performance of a specific function". SOPs will allow CRU to comply with the FDA's regulations.

CRU is also accountable to the FDA and CRU internal auditing group for following internal SOPs. SOPs are essential to a well run clinical trial. They indicate clinical researcher's (CRs) level of expertise and training in the clinical research arena. The SOPs available are very generic. Clinical researcher should customize them for the required business. Omit any procedure that does not apply to the business and add text to describe any procedure that CR do perform that is not included.

Work Instructions / Working Guidelines

SOPs should contain the high-level requirements to govern the processes. Clinical researchers will want to write work instructions or working guidelines to give more detail to the processes outlined in the SOPs. SOPs should not contain actual job titles and phone numbers as these may be outdated easily.

SOP contains attachments which include guidelines, checklists, logs and other forms. These files are not required (with just a few exceptions), but are available as a resource. The attachments may be used as is, modified or adapted as needed or simply replaced with forms that have been developed previously and are already being used. The sole exceptions are certain template forms (e.g. SOP template) which are to be used as written.

In conclusion, this SOP Manual is the starting line in establishing the institutional culture of research excellence, and not the finish line. These SOP must be read, understood and used by all staff for an appropriate period of time. After this "shake-out" period, the SOP Manual should be assessed adapted and improved as experience dictates. Thereafter, periodic review and updating will ensure this SOP Manual remains a living, useful document, and not a placeholder on a shelf, gathering dust.

Suggested Reading

1. Celine M Clive. Hand book of SOPs for Good Clinical Practise. Interpharm/ CRC publication.
2. Guidelines for writing SOP for clinical trials – Instruction manual. CDSCO publication.
3. O.H.Shah. SOP guidelines. Business Horizon's Pharmaceutical publishers. 2000.
4. Good clinical practice standard operating procedures for clinical researchers. Ed Josef Kolman, Paul Meng, Graeme Scott. Published by John Wiley and Sons 2001.

CHAPTER 23

Archiving Clinical Research Documents

M.U.R. Naidu and P. Usha Rani

3.1 Introduction

Definition: Archiving in the context of clinical research relates to the collection or long term storage of essential documents that individually and collectively ermit the evaluation of the conduct of a clinical research and the quality of ata produced. List of documents which are considered to be essential documents mentioned in section 8 of ICH-GCP. According to the principle of ICH-GCP, ie principal investigator is responsible for arranging the archival of their research ocuments.

Archiving has always been an important part of compliance with GCP. The uidelines only apply to clinical research that supports or will be used in a iarketing application including clinical research involving an investigational iedicinal product, whether conducted by commercial sponsor or non-commercial cademic institution. However, clinical trial directive does not apply to non-iterventional or observational studies. Additionally, there are other considerations) support this.

(a) The data may be needed by the sponsor to aid research into different dosage regimens, use for other indications or other therapeutic areas.

(b) The data may be needed to support marketing license requirements in other parts of the world.

(c) It may be necessary to do the re-analysis of the data to assist marketing departments or to answer any questions that may subsequently arise during clinical use.

(d) In cases of suspected medical fraud or other litigation, it is vital that the sponsor is able to produce all the relevant documentation

23.2 Stake Holders

Archivist: Person professionally educated, trained and experienced, responsible for the management and the administration of the archives. The archivist is responsible for, preserving the clinical documents, ensuring the integrity of the study and enabling and facilitating access for audit, legal and regulatory requirements. The archivist must be independent of the conduct of studies. If several people are employed within the archive they must work under the supervision and management of the nominated archivist.

Investigator: Person responsible for the conduct of the clinical trial at a trial site, including the maintenance of clinical records during and after the completion of the trial. If there is a team of individuals working on the trial at a given site, then the responsible team leader is called the principal investigator.

Sponsor: Individual or organization responsible for the conduct of a clinical trial. The sponsor is also responsible for ensuring that the completed study documentation is retained under environmental conditions that will preserve it for as long as necessary to comply with legal and regulatory requirements.

Study Manager: Person nominated by the sponsor or the Contract Research Organization to be responsible for the day to day conduct of the trial. The study manager is responsible for ensuring completeness of the trial master file (TMF) and its timely transfer to archive or sponsor.

Clinical Research Associate (CRA): They are research professionals carrying out activities that may include investigational site selection, set up, initiation, monitoring and close-out and can be involved in all operation aspects of the clinical phases of drug development.

Clinical trial Administrators: They administer, maintain and co-ordinate the logistical aspects of clinical trials according to GCP and relevant Standard Operating Procedures (SOPs) and act as a pivotal point of contact for the clinical trial team.

23.3 What Documents Must Be Archived

As per ICH-GCP, essential documents – those documents which individually and collectively permit the evaluation of conduct of a study and the quality of the data produced – are to be archived. Source documents (hard copies) must be archived and this includes – observation and source data contained in medical records, data collection forms, or case record forms and research related records held in supportive departments like radiology, pharmacy, laboratory medicine. If there are any electronic records of documents, same must be backed up and retained with the hard copies.

23.4 How long Should the Documents be Archived

The ICH-GCP Guidelines (section 5.5.11) states that essential documents are retained until atleast 2 years after the last approval of a marketing application in an ICH region or atleast 2 years have lapsed since the formal discontinuation of clinical development of an investigational product.

Further, Directive 91/507/EEC states that the investigator shall arrange for the retension of the patient identification codes for atleast 15 years after completion or discontinuation of the trial. Patient files and other source data shall be kept for the minimum period of time permitted by the institute or hospital. Retension schedule, including any actions to be taken prior to document destruction, should be agreed before commencement of the study. These documents should be retained for a longer period however if required by the regulators or by agreement with the sponsor. It is the responsibility of the sponsor to inform the investigator or institution when these documents no longer need to be retained.

The medicine for human use amendment regulation (clinical trial) 2006 states that archival of study master file is a requirement for all CTIMPs studies (Clinical trial of investigational medicinal product). These documents should be archived for at least 5 years after the conclusion after the trial.

23.5 When to Archive Trail Master File (TMF)

Neither ICH-GCP guidelines nor GCP Directive have specified that at what point during a clinical study, Trial master file (TMF) and related documents need to be archived. Thus there are two systems followed – (1) Pre-archiving or (2) Archiving everything at study completion. However, whatever system is followed, it is the responsibility of the study manager to ensure the TMF is complete in quality and contents and arrange for its timely transfer to archives.

5 A. Pre-archive system: Even as the study is going on, some pharmaceutical companies and sponsors have developed a system of pre-archiving study documents, which is also known as in-stream archiving, where, as and when the study documents are received in the clinical department, they are logged, scanned and then made available to the study team electronically. In a multinational, multicentric study, where all the documents are being centralized, working copy of documents can also be used while the original is retained in the TMF and stored in a dedicated archive area. This dedicated area is maintained by specialist staff, who are experienced in filing and records management. The working copy should be clearly identified and it should either be printed out on colored or suitably water-marked paper, or otherwise distinguished from the original.

Once the original documents have been transferred to the pre-archive they are not then released to the study staff again. Both the hard copy and the electronic version are filed according to the company's filing structure and SOPs. As the documents are logged it is very simple to maintain an index of essential documents and ensure the completeness of the study documentation. As all the documents are available electronically all staff, whether office or field based can access the documents and with global studies the documents can also be accessed in other countries. If a member of the study team needs a paper working copy of a document they can print it out for their own use and at the end of the trial simply shred it as the original is safely stored in a archive.

Advantages with the pre-archive system:

1. Mainly the original documents are stored in a secure storage area with access restricted to the archive staff, and so the original documents are not handled.
2. The other advantage is that all the files are maintained by specialist staff, ensuring a high standard of consistent and accurate filing. It is easier to flag up any missing documentation. The files are held in a format ready for archive at study end. On completion of the study, the TMF, which has already been logged has only to be moved from the pre-archive, i.e. current active filing section into the 'deep-archive', i.e. the inactive section.

Disadvantages with this system:

1. Mainly it requires a specific dedicated storage area with access limited to the archive staff and working area for these staff.
2. It also needs the specialized dedicated staff to work in the pre-archive, thus it is expensive to implement when considered against the alternative of study staff filing their own documents and storing them in the clinical research department.

5 B. **Archiving at study completion system:** Alternatively, the TMF can be filed by clinical trial administrators and CRAs with in the clinical research department. The documents are filed according to the company's filing structure and SOPs. While the study is ongoing, the essential documents are accessed as and when necessary.

Disadvantages with this system include:

1. The documents are liable to injury or misplaced or even lost if removed from the files.
2. Documents are not secure. When this system is adopted, data managers and statisticians will usually send their data to archive separately and often at a later date than the clinical team. This is not necessarily a problem but misunderstanding and oversights may arise. However, if the CRO or Sponsor has a secure document room for maintaining the TMFs, the security aspects can be resolved.

Advantages of the system are:

1. This system does not require so many staff for implementation.
2. Uses less space than the pre-archive.
3. Hence, it is a much cheaper alternative.

Finally, whatever system is being used, there should be proper SOPs in place specifying the procedure and clearly stating precisely who is responsible for the documentation at each stage of its life. The SOP must also stipulate the point in the process at which the TMF becomes the responsibility of the archivist especially with multinational trials, where documents are to be retained centrally and some documents need to be retained in the country of original to comply with local regulatory requirements. Further, if when the study is contracted out to a CRO and they hand the files back to the sponsor at the end of the trial, there should be a proper contract as to which documents should be held by the CRO and which by the Sponsor during the conduct of the trial.

23.6 The Archive Building

There are several important factors to consider when selecting a site for an archive. However few people have the opportunity to build the 'perfect' archive and usually the archive has to be a compromise between available space, accessibility an often the budgetary constraints. The following factors must be considered:

- Location
- The building itself, Internal furnishings, Environmental controls
- Security

6 A. Location: The archive should not be built in an area at particular risk from earthquakes, tidal waves, landslides, fuel storage depots in vicinity and even rodents and insects. Even if the archive has been built to withstand an earthquake and if all the buildings around it are damage, access the archive may not be possible. If the archives is based at the pharmaceutical company consideration should be given to a separate building away from any manufacturing activity.

Where no suitable on-site archives exist, investigators must arrange for documents to be archived in a dedicated off-site facility. Investigators must be provided with details of the requirements for archives and the same should be adhered to. Investigator should ensure that the archive facility meets the expected standards and ensure that the documents are not lost, damaged or prematurely destroyed.

6 B. Physical Requirements: Whether the building has been purpose built or as more usually it has been selected and adapted from an existing building, it must be constructed in such a way as to be whether proof and provide some degree of fire resistance. The walls and roofs should be constructed from a material with a high thermal capacity, such as brick, which is good insulating material. If the archive is within an existing building it should be a self-contained unit. The building must be large enough to those not only the current archive contents but allow for future deposits to the archive. If possible the basement should be avoided because of the risk of flooding and he presence of water and heating pipes. The water and heating pipes must not run through the archive because of the danger of leakage from them.

Doors should be constructed in such a way to ensure they are water proof and secure. Any external door should have excluders fitted to ensure they are whether proof and prevent rodents slipping in under the doors. There should be a covered area for the delivery and collection of archive boxes. It is preferable for security and environmental considerations for the archive not to have windows but if they are present then they should also be waterproof and secured in such a way as to prevent access.

It is preferable for the archive to be located on the ground floor both for convenient deliveries and also because of the weight of the storage of archive boxes. If the archive is located on an upper floor then a structural engineer will be needed to check the construction of the floor and calculate its load bearing capacity. It may well be necessary to reinforce the flooring.

The walls, doors and floors should provide a minimum of 4 hours fire resistance. Also any materials, paints etc. used within the facility should no be flammable or give off harmful gases. As far as possible electrical installation should be outside the archive and there should be no naked light bulbs with the facility. Further, the archive facility must be large enough to store the current contents of the archive and allow for future deposits to the archive.

The environmental conditions under which documentation is stored is increasingly important during long term storage. The ideal environmental conditions are a stable, ambient temperature 13-18^{0}C and relative humidity between 55-65%. Dust levels should be kept minimum.

Trial records also include data recorded on microfilm, floppy disks, tapes, x-ray films etc, hence the storage conditions should be appropriate for the records being stored. Broadly a temperature range of 13-18^{0}C and relative humidity of 50-60% is acceptable. However specific temperature requirements include:

Compact discs, computer tapes and other magnetic media	Temperature range : 18-20^{0}C Relative humidity : 35% - 45%
X-rays, Photographic films etc	Temperature range : 10-18^{0}C Relative humidity : 35% - 45%

6 C. Security: The security of the archive is paramount. The archive should have secure perimeter fencing, 24 hours security guards and closed circuit television. There should be intruder alarms fitted to all doors and windows, monitors on a 24 hour basis and should automatically alert the police to the intrusion. Access to the archive should be limited to archive staff by means of an entry key code or magnetic swipe card. A log must be maintained of all visitor and inspectors. When any building work or maintenance is undertaken the companies and staff involved should be carefully vetted and all workmen must sign the access log.

The other major security concern within the archive is fire. Hence, prevention of fire is better than relying on the procedures to contain a fire. Smoking must be absolutely forbidden in the archive and also attention must be paid to regular checking of electrical installations and wiring. There must be a fire detection system in place around the entire facility that both alerts staff to the fire and also automatically raises the alarm with the local fire brigade. There should be a central control panel for the system, in an area that is continually manned. A system based on smoke detection rather than heat or flame sensors is most suitable as archive boxes of paper tend to smolder and emit smoke rather than dramatically burst into flames. There are extremely sensitive systems available, known as VESDA (very early smoke detection apparatus), these are up to 200 times more sensitive than normal smoke detectors.

At times, sprinkler system fitted within the archive is used and there should be handheld fire extinguishers, for both paper and electrical fires, provided at various points throughout the archive facility. It is important that all this equipment is regularly inspected and maintained. However, it is fairly expensive to install and maintain so are really only practical in small facilities.

The best way to maintain temperature and humidity requirements is to install an air conditioning system. This can heat or cool the air, humidify or dehumidify the air, assist circulation of the air and also to clean pollution from the air. Ideally the system, once installed should be left running 24 hours a day. The system will also need to be regularly serviced and maintained. There must also be a process in place for monitoring that the system is maintaining the environment with in the required levels. The sensors should be placed around the archive to give a full record of the conditions, so for example near the roof, by the door and near an outside wall. There should be log to record when and where with in the archive the readings have been taken and a record of these readings. This does not have to be performed manually but can be controlled by a computerized system. If for any reason the temperature or humidity levels are outside of the criteria, a reason for the deviation should be recorded along with what action was taken to correct the problem.

Conclusion

Record keeping is crucial during and after the trial and determines the standard of work by the investigator. It also reflects compliance of sponsor and investigator with GCP. Storage, destruction and retrieval of documents and records must be discussed, documents and recorded before the start of the trial. Good record archival is therefore very important to success of a trial.

Suggested Reading

1. Archiving Clinical Trial documents by Elizabeth Hooper, Alec Deighton published by Institute of Clinical Research, 2006 ISBN 1905238126, 9781905238125, pg1-50.
2. EFGCP Update Paper: Good Clinical Practice and Computer Use at Investigator Sites.
3. BS5454:1989 British Standard Recommendations for the storage and exhibition of archival documents.

Example of Archive Label Format

Project title ____

SponsorName ____

Contact details ____

Protocol Title : ____

CT Number ____________________ IRB Number ________________

Principal Investigator ____

Date archived ___________ Date to be destroyed _________________

Box [insert number] of [insert total number] and

Brief description of contents ___________________________________

Example of Archiving Log

Sponsor Name	
Protocol Number	
Ethics Number	___/__
Protocol Title	
Investigator	
Directorate	

Please detail below the Case Record Form Numbers

Please indicate the number of archiving boxes, specific contents per box and exact storage location

	Contents of Box	**Storage Location**
Box 1		
Box 2		
Box 3		
Box 4		
Box 5		
Box 6		

- Archive storage labels should be put on each storage box·
- Any details regarding a change in storage during the archiving period should be documented·
- The Investigator Declaration should be completed·
- This log should be stored in a specified location by

Principal Investigator Declaration

- All essential documents are filed within the Investigator File as per the requirements of ICH-GCP ☐
- All consent forms are in the Investigator File in numerical order ☐
- The archive boxes will be stored as per requirements to conform to ICH-GCP ☐

Principal Investigator signature..

Date:..

CHAPTER 24

Evidence - Based Medicine

M.U.R. Naidu and P. Usha Rani

24.1 Introduction and Overview

Evidence-based medicine (EBM) aims to apply uniform standards of evidence gained from the scientific methods to certain aspects of medical practice. According to the Centre for Evidence-Based Medicine *"Evidence-based medicine is the conscientious, explicit and judicious use of current best evidence in making decisions about the care of individual patients.*

EBM recognizes, that many aspects of medical care depend on individual factors such as quality and value-of-life judgments, which are only partially subject to scientific methods. Specifically, EBM seeks to assess the quality of evidence relevant to the risks and benefits of treatments (including lack of treatment), Practicing evidence-based medicine requires clinical expertise, but also expertise in retrieving, interpreting, and applying the results of scientific studies and in communicating the risks and benefits of different courses of action to patients, given the rapidly accumulating body of research that constitutes EBM.

There are three distinct, but interdependent, areas of EBM :

1. First is to treat individual patients with acute or chronic diseases by treatments supported in the most scientifically valid medical literature.

Thus, medical practitioners would select treatment options for specific cases based on the best research for each patient they treat.

2. The second area is the systematic review of medical literature to evaluate the best studies on specific topics. This process can be very human-centered, as in a journal club, or highly technical, using computer programs and information techniques such as data mining. Increased use of information technology turns large volumes of information into practical guides.
3. Finally, evidence-based medicine is a medical "movement" which advocates work to popularize the method and usefulness of the practice in the public, patient communities, educational institutions, and continuing education of practicing professionals.

24.2 History

Since ancient times, testing medical interventions for efficacy has existed - Avicenna's *The Canon of Medicine* in the 11th century. However in the 20th century Professor Archie Cochrane, a Scottish epidemiologist, through his book *Effectiveness and Efficiency: Random Reflections on Health Services* (1972) and subsequent advocacy, caused increasing acceptance of the concepts behind evidence-based practice. Cochrane's work was honored through the naming of centres of evidence-based medical research — *Cochrane Centres* — and an international organization, the Cochrane Collaboration. The explicit methodologies used to determine "best evidence" were largely established by the McMaster University research group led by David Sackett and Gordon Guyatt. The term "evidence based" was first used in 1990 by David Eddy The term "evidence-based medicine" first appeared in the medical literature in 1992 in a paper by Guyatt.

24.3 Types of EBM

Two types of evidence-based medicine have been proposed.

(a) **Evidence-based guidelines:** Evidence-based guidelines (EBG) is the practice of evidence-based medicine at the organizational or institutional level. This includes the production of guidelines, policy, and regulations.

(b) **Evidence-based individual decision making:** Evidence-based individual decision (EBID) making is evidence-based medicine as practiced by the individual health care provider. There is concern that current evidence-based medicine focuses excessively on EBID.

There are variable reports about whether evidence-based medicine is effective. Using the classification scheme above —dividing evidence-based medicine into evidence-based guidelines (EBG) and evidence-based individual decision (EBID)— may explain the conflict. It is difficult to find evidence that EBID improves health care, whereas there is growing evidence of improvements in the efficacy of health care when evidence-based medicine is practiced at the organizational level.

24.4 Categories of Evidence

According to the strength of freedom from the various biases that beset medical research, EBM is categorized into different types of clinical evidence and then ranked. *For example, the strongest evidence for therapeutic interventions is provided by systematic review of randomized, double-blind, placebo-controlled trials involving a homogeneous patient population and medical condition.*

In contrast, patient testimonials, case reports, and even expert opinion have little value due to the biases inherent in observation and reporting of the studies.

Various systems to stratify evidence by quality have been developed, most popular of which is one developed by the U.S. Preventive Services Task Force for ranking evidence about the effectiveness of treatments or screening:

- Level I: Evidence obtained from at least one properly designed randomized controlled trial.
- Level II-1: Evidence obtained from well-designed controlled trials without randomization.
- Level II-2: Evidence obtained from well-designed cohort or case-control analytic studies, preferably from more than one center or research group.
- Level II-3: Evidence obtained from multiple time series with or without the intervention. Dramatic results in uncontrolled trials might also be regarded as this type of evidence.
- Level III: Opinions of respected authorities, based on clinical experience, descriptive studies, or reports of expert committees.

The UK National Health Service uses a similar system with categories labeled A, B, C, and D. The above Levels are only appropriate for treatment or interventions; different types of research are required for assessing diagnostic accuracy or natural history and prognosis, and hence different "levels" are required.

- Level A: Consistent Randomised Controlled Clinical Trial, Cohort Study, All or None, Clinical Decision Rule validated in different populations.
- Level B: Consistent Retrospective Cohort, Exploratory Cohort, Ecological Study, Outcomes Research, Case-Control Study; or extrapolations from level A studies.

- Level C: Case-series Study or extrapolations from level B studies
- Level D: Expert opinion without explicit critical appraisal, or based on physiology, bench research or first principles

Oxford Centre for Evidence-based Medicine suggests levels of evidence (LOE) according to the study designs and critical appraisal of prevention, diagnosis, prognosis, therapy, and harm studies.

A newer system that takes into account more dimensions than just the quality of medical evidence and makes "Extrapolations" where data is used in a situation which has potentially clinically important differences than the original study situation , is suggested by the Grade Working Group. Thus, the quality of evidence to support a clinical decision is a combination of the quality of research data and the clinical 'directness' of the data.

Though there are subtle differences between systems and the individual studies still require careful critical appraisal, the purpose is the same: to guide users of clinical research information about which studies are likely to be most valid.

24.5 Categories of Recommendations

Recommendation for a clinical service is formulated on the balance of risk versus benefit of the service *and* the level of evidence on which this information is based. The U.S. Preventive Services Task Force uses the following categories:

- Level A: Good scientific evidence suggests that the benefits of the clinical service substantially outweighs the potential risks. Clinicians should discuss the service with eligible patients.
- Level B: At least fair scientific evidence suggests that the benefits of the clinical service outweighs the potential risks. Clinicians should discuss the service with eligible patients.
- Level C: At least fair scientific evidence suggests that there are benefits provided by the clinical service, but the balance between benefits and risks are too close for making general recommendations. Clinicians need not offer it unless there are individual considerations.
- Level D: At least fair scientific evidence suggests that the risks of the clinical service outweighs potential benefits. Clinicians should not routinely offer the service to asymptomatic patients.
- Level I: Scientific evidence is lacking, of poor quality, or conflicting, such that the risk versus benefit balance cannot be assessed. Clinicians should help patients understand the uncertainty surrounding the clinical service.

24.6 Statistical Measures

Evidence-based medicine attempts to express clinical benefits of tests and treatments using mathematical methods. Tools used by practitioners of evidence-based medicine include:

- Likelihood ratios. The pretest probability of a particular diagnosis, multiplied by the likelihood ratio, determines the posttest probability. This reflects Bayes' theorem. The differences in likelihood ratio between clinical tests can be used to prioritize clinical tests according to their usefulness in a given clinical situation.
- The area under the receiver operator characteristic curve (AUC-ROC) reflects the relationship between sensitivity and specificity for a given test. High-quality tests will have an AUC-ROC approaching 1, and high-quality publications about clinical tests will provide information about the AUC-ROC. Cutoff values for positive and negative tests can influence specificity and sensitivity, but they do not affect AUC-ROC.
- Number needed to treat or Number needed to harm are ways of expressing the effectiveness and safety of an intervention in a way that is clinically meaningful. In general, NNT is always computed with respect to two treatments A and B, where for example, A is a 5-year treatment with a typical drug and B is a placebo and the endpoint defined as the appearance of colon cancer in the 5 year period. If the probabilities pA and pB of this endpoint under treatments A and B, respectively, are known, then the NNT is computed as 1/(pB-pA). The NNT for breast mammography is 1/285, so 285 mammograms need to be performed to diagnose one breast cancer.

24.7 Quality of Clinical Trial Publications

Evidence-based medicine attempts to objectively evaluate the quality of clinical research by critically assessing techniques reported by researchers in their publications.

- Trial design considerations. High-quality studies have clearly-defined eligibility criteria, and have minimal missing data.
- Generalizability considerations. Studies may only be applicable to narrowly-defined patient populations, and may not be generalizable to clinical practice.
- Followup. Sufficient time for defined outcomes to occur can influence the study outcomes and the statistical power of a study to detect differences between a treatment and control arm.
- Power. This is the number of patients enrolled in the trial. A complex mathematical calculation can determine if the number of patients is

sufficient to detect a difference between treatment arms. A negative study may reflect a lack of benefit, or simply a lack of sufficient quantities of patients to detect a difference.

24.8 Limitations of Available Evidence

It is recognised that not all evidence is made accessible, this can limit the effectiveness of any approach, and that effort to reduce various publication and retrieval biases is required.

Failure to publish negative trials is the most obvious gap, and moves to register all trials at the outset, and then to pursue their results, are underway. Changes in publication methods, particularly related to the Web, should reduce the difficulty of obtaining publication for a paper on a trial that concludes it did not prove anything new, including its starting hypothesis.

Treatment effectiveness reported from clinical studies may be higher than that achieved in later routine clinical practice due to the closer patient monitoring during trials that leads to much higher compliance rates.

24.9 Criticism of Evidence-Based Medicine

Critics of EBM say lack of evidence and lack of benefit are not the same, and that the more data are pooled and aggregated, the more difficult it is to compare the patients in the studies with the patient in front of the doctor — that is, EBM applies to populations, not necessarily to individuals. There are several areas of disagreement between EBM supporters and detractors as well as unanswered questions about the role of EBM in modern healthcare. *The limits of evidence-based medicine*, Tonelli argues that "the knowledge gained from clinical research does not directly answer the primary clinical question of what is best for the patient at hand." Tonelli suggests that proponents of evidence-based medicine discount the value of clinical experience. Although evidence-based medicine is becoming regarded as the "gold standard" for clinical practice and treatment guidelines, there are a number of reasons why most current medical and surgical practices do not have a strong literature base supporting them.

1. The first type of criticism involves the philosophical underpinnings of EBM, which is based on empiricism. EBM elevates experimental evidence to primary importance over pathophysiological and other forms of knowledge, and implicitly assumes that scientific observations can be made independent of the theories and biases of the observer.
2. The second theme is that the definition of evidence within EBM is narrow and excludes information important to clinicians. EBM grades evidence according to the methods used to collect it. However, EBM does not provide a means to integrate other, non statistical, forms of medical information, such as professional experience and patient specific factors.

3. Third, EBM is not 'evidence-based' because it does not meet its own empirical tests for efficacy. Though, EBM proposes that patient care can be improved by basing clinical decision-making on information from statistically valid clinical trials, ironically, there is no evidence (as defined by EBM) that this is actually the case.
4. Fourth, the usefulness of applying EBM to individual patients is limited. Clinicians must balance general rules, empirical data, theory, principles, and patient values and apply them to individual people. This requires a great deal of clinical judgment.
5. Lastly, EBM has been criticised for reducing the autonomy of the doctor-patient relationship by limiting the patient's right to choose what is best in their individual circumstances. EBM could be used as a cost- cutting tool to deny treatment where interventions are not 'proven' effective. On the other hand, EBM could also increase costs by 'proving' the efficacy of some expensive interventions. Currently, the net effect of EBM is unknown

Suggested Reading

1. Fitzpatrick, Michael (2008). "Taking a political placebo". Spiked Online.
2. Sackett DL, Rosenberg W, Gray JA, et al. Evidence based medicine: what is it and what it isn't. BMJ. 1996; 312:71-2.
3. Sackett DL, Straus S, Richardson S, Rosenberg W, Haynes B. Evidence based medicine: how to practice and teach EBM. 2nd ed. London: Churchill Livingston.
4. Paul Glasziou, Jeffery Aronson, Les Irwig Evidence based medicinal monitoring. Wiley-Blackwell 2008.

CHAPTER 25

Clinical Research Data Management

P. Usha Rani and M.U.R. Naidu

25.1 Introduction

Statistical considerations must be an integral part of study design and the description and analysis of the final results. The aim should be to integrate the clinical and statistical aspects as completely as possible through all phases of a clinical study. Statisticians and data managers should have advisory and operative functions in the design of clinical studies, in reviewing study protocols, CRFs and manuscripts, and in the design of computer systems for data processing. The operative functions consist of writing or advising on statistical sections in protocols and reports, adapting or developing appropriate methods for the statistical analysis, performing the analysis and making the statistical interpretation of the results. The protocol and the final study report should be reviewed and commented upon by a statistician.

25.2 Protocol

When preparing a study protocol, statistical considerations form an integral part and hence special attention should be paid to the following aspects:

- that the experimental design of the study is appropriate to the purpose of the study and the nature of the study condition;
- that the sample size is large enough to allow detection of clinically meaningful differences or to document clinical equivalence as appropriate, in variables specifically related to the hypotheses stated in the protocol;
- that the calculation of the sample size is based on sufficient statistical power and adequate levels of significance specified before commencement of the study;
- that the calculation of sample size takes into account any planned interim or sub-group analyses;
- that the randomisation procedures must be scientifically justified in the protocol;
- that for comparative, open-label studies measures taken to avoid bias are specified (preferably assigned by central randomisation);
- that the reasons and procedures for any planned interim analyses, whether used in sequential design for possible early termination or for any other purpose, are clearly specified in writing. It is advisable to have an independent supervisory committee to perform interim analyses to maintain study blindness among investigators and others involved;
- that the general approach to the statistical analysis is described, including methods of adjusting for possible imbalance in prognostic factors at baseline;
- that criteria which can be foreseen for including and/or removing participants from certain analyses are defined;
- that the use of intention-to-treat analysis and/or per protocol analysis is specified.
- that methods for dealing with/analyzing incomplete data (missing observations, premature withdrawals, etc.) are specified; and
- that specific statistical models and methods which can be foreseen are used even if some flexibility must be allowed at the time of analysis.

25.3 Randomisation

Detailed description and documentation of the procedure for the randomisation of participants is necessary in a clinical research study. In a blinded, randomised study it is usually necessary to supply and keep the treatment code for each individual participant at both the study site and with the sponsor. Further, the methods of unblinding the treatment code and its documentation must be specified in the protocol. Breaking the treatment code must be possible 24 hours a day in case of an emergency and the procedures and persons involved clearly designated in the protocol and related documents. The date, reasons for, and name of the individual breaking the study code must be documented. Before the treatment code is broken for statistical analysis, the code for each participant must be returned to the sponsor with a documented explanation for each episode where the code was broken. Any master code supplied (e.g. to the pharmacy) must be returned to the sponsor. No copies of the code should be taken by any person involved in the study. Copies of the treatment code will be available to the investigator at the end of the study after the database is 'locked'. (For more details refer to chapter on Randomisation)

25.4 Data Management

Data management aims at transferring the information from the participant, efficiently and without errors, into data in the statistical database. All steps involved in data management should be part of a standard operating procedure and should be documented to allow for a step-by-step retrospective assessment of data quality and study performance; i.e. an audit trail. Documentation is facilitated by the use of such means as checklists and forms giving details of action taken, dates and the individuals responsible.

25.4.1 Data Confidentiality and Transfer

An important aspect of the securing data integrity is the safeguarding of "blindness" with regard to treatment assignment. It starts with the randomization of participants into treatment groups and it is maintained through all steps of data processing up until the decision to break the code is formally taken.

The confidentiality of the computer aided database must be secured by appropriate standard operating procedures including passwords for all staff involved. The use of a computer system which identifies and records all information on individuals who had access to the system, and dates all changes to the information as an audit trail is recommended. Satisfactory maintenance and back-up procedures for computer databases must be provided.

25.4.2 Case Report Form (CRF)

The case report form should be so designed to capture data on all the aspects of the trial as per the protocol requirement. Further, there should be consistency in the use of reference codes, terminology and format in the case record form. Standardization in the design of forms and computer programmes used in the data processing and statistical analysis will save time and prevent errors.

25.4.3 Data Quality Control

The aim of quality control procedures is to minimise the effects of missing and inaccurate data. There should be standard operating procedures for data entry. Any changes made after data entry should be signed and dated by the authorized person. An audit trail should be available to trace the nature of any changes to data, the dates of changes and the person responsible for the changes. Further, data collection and entry should be performed continuously during the course of the study. It should be checked either by double-entry (preferable) or by proof-reading for the primary variables and on a random basis for other parameters. Continuous checking of each individual's data during the study can be aided by the use of appropriate descriptive statistics on each important variable in the database which (performed without breaking the code) is useful in the detection of doubtful and/or unusual data.

25.4.4 Code Breaking for Data Analysis

The data base is finally formally "locked" once the validation and editing process is concluded and the same is documented. Further, the treatment code can be broken and included in the data base for each individual participant. Data for each individual participant should be classified and coded with respect to its inclusion in the various statistical analyses planned in the study and the code entered into the data base.

25.5 Final Study Report

The protocol, statistical and clinical aspects should be integrated in order to obtain a final study report that is entirely consistent with the study data generated. Essential elements in the presentation of the results of a study of an investigational product are:

- baseline comparisons between the treatment groups;
- the number of participants actually randomised into the study by treatment group and the number of participants excluded from any of the analyses, by reason and by treatment group;
- major efficacy and safety results by treatment group in the form of tables, graphs, test variables and statistical parameters (e.g. p-values) as appropriate;
- an assessment of between-group differences with confidence intervals; and
- in multicentre studies, an evaluation of centre effect may be a valuable addition and should always be conducted where significant inter-centre variation is suspected.

An account must be made of missing, unused or spurious data during statistical analyses. All omissions of this type must be documented to enable review to be performed. The final report is then submitted to the Drug Regulator Authority.

25.6 Retension and Securing of Records

Both the principal investigator and the sponsor are obliged to retain records and data from the study for safety reasons and for audit and inspection subsequent to study completion. Acceptable archiving practices are 5 years for biological samples and 10 years for written materials. Samples, documents and any computer records should be retained in a secure place to prevent unauthorized access, loss or tampering.

25.7 Archiving by the Principle Investigator

Archiving of data needs to be performed as per the prevailing SOPs. The principal investigator is responsible for maintaining copies of all documentation which contains identifying source data and other essential documentation including, the study protocol and amendments, applications to the ethics committee, serious adverse event reports and all other correspondence relating to the study. All correspondence between the principal investigator and the sponsor must be preserved and be available on the request of the study sponsor, the regulatory authority, independent auditors or ethics committee. Investigational product accountability records detailing the storage and use of the product should also be retained. Adequate steps must be taken to ensure that the hospital case records of all participants in clinical research are retained for this period, which is longer than the time to destruction interval in some hospitals and institutions. In case of the principal investigator expressing inability to archive of the study documents and samples, the sponsor should be informed in writing about the location of the records and the name of the person responsible for their retention.

25.8 Archiving by the Sponsor

Similarly, the sponsor should retain copies of all essential documentation relating to the study which do not contain participant identifying information. These include reports to the regulatory authority, records of monitor-investigator contacts and investigational product supplies. The archiving should be done in a termite proof, fire and water proof secure storage facility accessible only to delegated personnel. Computer records must be reproduced in hard copy, which are to be signed and dated as a verified accurate copy of the original data. The verified hard copy should then be stored with other paper-based records, to overcome the possibility of loss or inability to read the information due to technological redundancy. Further, Computerized CD data should be maintained in the proper temperature controlled room. The same must be documented and secured by authorized staff.

Suggested Reading

1. Rondel R, Varley S, Webb C. Clinical data management. New York, NY: John Wiley and Sons Ltd. 2000

2. Colleen M Cox. Planning the data management process for a clinical trial, technology and data management. Monitor, September 2005.
3. Susanne Prokscha. Practical guide to clinical data management. CRC Press LLC 1999.

Chapter 24

Clinical Biostatistics

(a) Biostatistics for Clinical Researchers

P.V.N. Sridevi, M.U.R.Naidu and P.Usha Rani

26(a).1 Introduction

Clinical trial design, analysis and interpretation rely upon statistical principles. All statistical tests make certain pre-specified assumptions about the data. For example, parametric statistics rely on the assumption that data have a 'normal' distribution.

A necessary companion to a well designed clinical trial is its appropriate statistical analysis. Assuming that a clinical trial will produce data that could reveal differences in effects between two or more interventions, statistical analyses are used to determine whether such differences are real or are due to chance. An understanding of these assumptions is important not only for the design and analysis of clinical trials, but also for the critical appraisal of completed trials.

26(a).2 Descriptive Statistics

Descriptive statistics are commonly used to present the baseline characteristics of a population enrolled in a clinical trial. It includes tabulation of results, calculation of averaged (means), and minimum/maximum values (range) or standard deviations (spread). The mode is the most frequent observation, and the median is the middle of all the values when these are ranked in order. Descriptive statistics are an essential part of reporting the results of a study, but they do not provide comparisons within the data set, or between one data set and another.

Normal Distribution Curve

When studying continuous variables like blood pressure in a given study population, their distribution conforms to a bell shaped curve or 'normal' distribution typical for common, continuous biological measurements. The curve has particular mathematical properties. expressed by the equation. The normal distribution was described by the German mathematician Gauss (1777-1855), and is also called the 'Gaussian" distribution. The standard deviation helps to describe the spread of the observations. In a normal distribution, about 68% of observations lie within 1 (one) standard deviation either side of the mean, and about 95% of all observation lie within 2 (two) standard deviation either side of the mean. Other distributions, which are skewed to one side or the other, are also possible. These distributions are not 'normal' and in these type of data the "median" is usually a more useful parameter than the mean in representing the central location of these distributions.

26(a).3 Null Hypothesis and Alternative Hypothesis

To explain the null and alternative hypothesis, we can use an example of a study investigating a new treatment to reduce the mortality (death) rate after myocardial infarction. The simplest form of comparison is a two group placebo-controlled randomized trial. One group receives the active treatment and the other receives a placebo. The traditional 'null' hypothesis (H^0) for this trial assumes that the effect of the active treatment on mortality after myocardial infarction is the same as that of the placebo. The 'alternative' hypothesis (H^1) assumes that there is difference in the mortality rates between the active and placebo treatments.

The results of the trial above may show

- There is an apparent difference in mortality rates between the active and placebo treatment groups. In this case the null hypothesis (H^0) is rejected, and the alternative hypothesis (H^1) is accepted.
- There is apparently no difference in treatment effect between the two groups. In this case the null hypothesis (H^0) is accepted.

When making statistical comparisons between two groups of the trial, it is possible that there may be error associated with these calculations, and the effect seen may not completely reflect the true effect.

Types of error in clinical research

Alpha (Type I) and beta (Type II) errors.

- The alpha (type I) error occurs when the study finds a difference, but in truth non exists (a false positive). In this scenario the study findings support the alternative hypothesis and reject the null hypothesis, but in truth the null hypothesis is correct. Usually the acceptable alpha error is set at 0.05, or 1 in 20. In other words, the likelihood of incorrectly accepting H^1 is 1 in 20 or less.
- The beta (type II) error occurs when the study finds no difference, but in truth a difference does exist (a false negative). In this scenario the study findings support the null hypothesis, but in truth the alternative hypothesis is correct. The usual acceptable beta error is set at 0.8 that is if a 'true' difference was present it would be detected 4 times out of 5.

These errors are set at different levels because investigators have traditionally, and understandably, been more cautious about declaring false positives than false negatives. The stringency of each error can be varied according to circumstances. In some situations it may be more useful to assume that the alpha error is 0.01 or 1 in 100 and the beta error is 0.1, or 1 in 10. These variations have an impact on sample size estimations as discussed below.

In addition to the proposed alpha and beta errors, two further pieces of information are required to estimate the sample size: first, an estimate of the expected event rate in the control group of the trial (the group that is receiving standard care or placebo) is needed; second, an estimate of the likely reduction in event rates assuming the new treatment were to have a beneficial effect (for example, a 20% reduction in mortality). These estimates are usually based on previous studies (trials or registries), experience and some guesswork. Sometimes a range of treatment differences may be used. For example if only one previous small study is available, using the range of the mean effect from that study ('standard error of the mean') may be more useful in determining the correct sample size (say 10-30% reduction). One can understand that estimating sample size can be an inexact science! Nevertheless a carefully constructed sample size estimate is important for clinical trials, especially for large trials where the sample size has important effects on the resources required.

A common term used when describing the sample size of a trial is 'power'. Power refers to the ability of a particular sample size to detect the 'true' difference. Power is determined from 1 minus the beta error and is usually set at 0.8 (or 80%). To increase the power of a trial more patients would be required.

It should be noted that sample size is not a linear function. Thus, to detect a proportional difference of 15 % will take approximately four times as many patients as to detect a difference of 30% if all other assumptions remain equal. In large clinical trials the alpha and/or beta errors are often more stringent than the standard 0.05 and 0.2. A range of sample sizes based on variations of the errors and expected probabilities, should be considered for most trials and balance struck between power and available resources.

26(a).4 Classifications of Data

Results obtained in a clinical trial can be expressed in different ways. Measurement of blood pressure, weight, height, plasma cholesterol, serum creatinine and left ventricular ejection fraction are examples of ' continuous' variables. Continuous variables can be expressed at different levels of precision depending on the scale of the measurement methods (for example centimeters or millimeters). In contrast, measurement of death, hospital admission, strokes and myocardial infarctions are examples of 'dichotomous' or 'binary' variables. Dichotomous variables (also called 'categorical' variable) have only two possible outcomes. Thus, in a trial evaluating the effects of a new treatment on mortality, patients can be classified as having one of two conditions: either dead or alive. Most measurements in clinical trials can be classified as continuous or dichotomous.

Other ways of expressing measurements include assigning grades or 'ranks'. For example, when asked 'Do you feel depressed ?' a patient may be asked to select the most appropriate answer from the following: 'never, sometimes, often, always'. The patient's response has four possible categories that can be ranked from mild to severe. Statistical assumptions for these data are different from those for numerical data. An alternative approach to the reporting of continuous variables is to convert the continuous data into categorical data. This is done by dividing the patients into subgroups of either equal size or by chosen limits for each subgroup. The types of data recorded will determine the appropriate statistical tests required.

Comparison of continuous variables

The most commonly used statistical test for the comparison of results (continuous variables) from two sample populations is the 't' test. The 't' test compares the means of two sets of continuous variables and expresses the probability that any differences are due to the play of chance (i.e., that the null hypothesis is supported), or may be 'real'(i.e., the alternatives hypothesis is supported). This probability is often calculated from a 't' value.

Statistical approaches to comparistons of dichotomous variables:

The most common statistical approaches for the randomized comparison of two groups, using a dichotomous outcome such as death, are the chi-squared (X^2) test and the Fisher exact test. The chi-squared test involves determining the expected number of deaths in the treatment and placebo arms and then comparing these to the observed number. The test statistic used to assess these differences in death rate can be expressed as

$$X^2 = \Sigma \frac{(O - E)^2}{E}$$

Where O represents the observed frequencies and E the expected frequencies in each treatment arm. Under the null hypothesis of no difference, chi-squared has a distribution with approximately 1 degree of freedom. The Fisher exact test consists of evaluation of the sum of probabilities associated with the observed frequency table and all possible two-by-two tables that have the same row and column total as the observed data. When sample size is large, a test of the difference between two proportions also uses a normally distributed test statistic Z, which can be easily calculated by hand.

26(a).5 Test of Hypothesis with the Correct Statistical Tests

Early in the design phase, it is important that the investigators anticipate the statistical tests that are to be applied to the final results. Even though the statistical process will ultimately prove more extensive, the basic statistical approach should be thought early in the design.

Although the art of statistical testing is often seen as a mystical, and sometimes unfathomable process for new investigators, the basic approach for choosing the right statistical test is really quite simple. To choose the right statistical test in the majority of circumstances, the investigator must ask four simple questions .The answers to these four questions will determine the most appropriate statistical test in the majority of instances.

1. One asks, "What variables are being compared?" To answer this question, the investigator identifies the independent variable and the outcome variable that are to be compared in the central hypothesis test at the end of study.
2. When the variables are identified, the investigator asks, "What is the form of these variables?" and identifies the type of each variable. Is each a categorical, dichotomous, multicategoried, ordered variable, or a continuous (or interval) variable?"
3. If the outcome variable is continuous, the investigator asks, "Does it meet the normality assumption?" This question is answered by graphing the distribution of the outcome variable in the population to be studied. This

can often be done by data obtained on this or a similar population from previous studies or from the literature or the investigator may wish to perform a pilot study to estimate the distribution.

4. The investigator asks, "Are the comparison groups independent or paired?". In simple experiments, most cohort studies, and some case-control studies, the observations on the study subjects are independent of one another, and the answer to the question is "No". In repeated measures, experimental designs and in matched case-control studies, the observations on the subjects are paired, in which case the answer to the question is "yes".

26(a).6 Summary

The four practical questions to ask in determining the most appropriate method of statistical analysis and test of statistical significance are:

1. What are the independent and dependent variables?
2. What type of variables are they?
   Categorical, dichotomous
   Categorical, multicategoried, nominal
   Categorical, multicategoried, ordered
   Continuous (or interval)
3. If the outcome variable is continuous, does it meet the Normality Assumption
4. Are the observations independent or paired
   (a) The answers to these questions will allow selection of appropriate statistical tests in the majority of instances, but confidence in the selection requires ample research experience or consultation with the statistician.
   (b) Paired observations are most often found in repeated measures experiments and matched case-control studies.

The tables given below indicate the usual analysis strategy, the measure of association used, and the test of statistical significance that is most commonly used and also answers these four above questions. In the majority of clinical studies, the outcome variable is either a dichotomous (two category) categorical variable (Table 26.1) or a continuous (or interval) variable (Table 26.2). In each of these instances, the independent variable can be any of the four types of variables, and these are portrayed in the first column in Table. For each of the resulting rows of the two tables, the usual type of statistical analysis performed, the most useful of the association between the independent and dependent variables, and the usual tests of statistical significance (to estimate the probability of a Type 1 sampling error) are given.

Table 26.1 Simplified scheme for choosing statistical tests for two category (dichotomous) outcome variables.

Group Significance variable	Outcome variable	Measure of analysis	Test of statistical association	Nonparametric procedure
Two category (dichotomous)	Two-category	2x2 contingency table	Relative risk (odds ratio)	Chi-square test Fisher's exact test.95% confidence interval of the relative risk McNemar's test
Multi-category Nominal	Two-category	nx2 contingency table	Uncertain Coefficient	Chi-square test
Multi-category, ordered	Two-category	Nx2contingency table	Goodman Kruskalgamma	Chi-square test Chi-square test For trend
Continuous	Two-category	Usually recode independent variable to categories and analyze it by the appropriate method above, or use logistic regression analysis	Same	Same

For example, most commonly in cohort and case-control studies, both the independent and dependent variables are dichotomous (two categoried) categorical variables. Therefore, as per Table 1, which deals with dichotomous categorical dependent variables, the appropriate analysis, shown in the first row, is a 2x2 contingency table, the relative risk is the most appropriate measure of association between the two variables, and the usual significance test is a p value calculated with chi-square test or Fisher's exact test. Increasingly, the 95% confidence intervals of the relative risk are being used in the place of the p value from the chi-square test of Fisher's exact test. If the observations were paired, as in a repeated measures experimental design or a matched case-control study, McNemar's test is used instead of the chi-square or Fisher's exact test.

When the dependent variable is continuous (or interval) data, then Table 26.2 portrays the options for continuous outcome variables. The most appropriate analyses must also consider whether the continuous outcome variable is normally distributed or not, indicated in the third column in the table. If the continuous outcome variable is normally distributed, the investigator can use a powerful set of tests that compare groups. (Tests that compare the means of continuous variables are referred to by the cryptic terms as parametric tests). The measure of the association between the two groups is usually the difference in the means, and the most commonly used statistical test is student's t test.

If the variables are not continuously distributed, the distribution of the continuous outcome variable is skewed to the right or to the left, then the investigator has two choices. One can transform the outcome variable by an appropriate transformation (e.g. taking the logarithm, the square, of the square root of each patient's value) into a nearly normal distribution and then use one of the powerful sets of parametric tests that compare the means. Alternatively, one can use a different set of statistical tests that do not require the data to be normally distributed. These are referred to as nonparametric tests eg Wilcoxon's rank-sum test, or the MannWhitney.

Conclusions

Statistical assumptions and methods are essential in the design, conduct and analysis of clinical trials. Analysis plans summarizing the outcome measures and how they will be compared should be an integral part of all study protocols. This plan should be expanded and finalized prior to analyzing the data. Randomized studies require substantial resources, and it makes sense to obtain the best advice before using these resources. The advice of a statistician with experience in medical research should be obtained from the design stage through to the completion and analysis of all trials.

Suggested Reading

1. Hill AB. Principles of medical statistics. 9ed, London: The Lancet 1971
2. Anello C. Statistical issues in Comparative Clinical Clinical Trials. Drug Inf. 1983; 17:55
3. Armitage P. Statistical methods in medical research. New York; Willy and Sons 1973.
4. Buncher CR and Tsay JY. Statistics in Pharmaceutical Industry. NewYork: Marcel Dekkar 1981.

Table 26.2 Simplified scheme for choosing statistical tests for continuous outcome variables.

		Parametric Procedures			Non-parametric
Group/predictor variable	Normality assumption satisfied	Analysis	Measure of association	Test of statistical significance	Test of statistical significance
Two category (dichotomous)	Yes	Compare means	Difference between means	Student t test / paired t test	Not needed
	No	Transform data to compare means	Difference between transformed means	Student t test / paired t test	Not needed
		Or use Non-parametric test			Mann Whitney U test / Rank sum test / Median test / Signed rank test
Multicategory Nominal	Yes	Compare means by ANOVA	Difference between means	F test	Not needed
	No	Transform the data & compare Means by ANOVA	Difference between transformed means	F test	Not needed
		Or Use Non-parametric test			Kruskal-Wallis Non-parametric ANOVA
Mulicategory Ordered	Yes	Compare Means by ANOVA	Difference between means	F test	Not needed

		Parametric Procedures	Non-parametric		
	No	Transform the data & compare Means by ANOVA	Difference between transformed means	F test	Not needed
		Or Use Non-parametric test			Kruskal-Wallis Non-parametric ANOVA
Continuous	Yes	Correlation or Regression Analysis	Pearson's r Or Regression coefficient	F test	Not needed
	No	Transform the data and do Correlation or Regression	Difference between transformed means	F test	Not needed
		Or Use Non-parametric test	Spearman's r Medians		Spearman's Rank order correlation

(b) Terminology used in Biostatics

M.U.R.Naidu, P.Usha Rani, P.V.N.Sridevi

6(b).1 Introduction to Biostatistics

iostatistics is the term used when tools of statistics are applied to the data that derived from biological sciences such as medicine.

Clinical trial design, analysis and interpretation rely upon statistical principles. ll statistical tests make certain pre-specified assumptions about the data. For xample, parametric statistics rely on the assumption that data have a 'normal' istribution. An understanding of these assumptions is important not only for the esign and analysis of clinical trials, but also for the critical appraisal of completed ials. Since, most studies involve the measurement of several characteristics ther than just one, multivariate methods may be required for appropriate analysis f results.

Statistics methods are utilized for several purposes: descriptive, relational nd comparative. Any Statistical Analysis is mainly divided into two parts. The rst part includes the organization and summarization of data and the second art includes reaching decisions about a large body of data by examining only a nall part of it. The concepts and methods necessary for achieving the first bjective are called **Descriptive Statistics,** and the second objective is reached rough the study of what is called **Inferential Statistics**.

6(b).2 Terminology used in Biostatistics

6(b).2.1 Biostatistics

he tools of statistics are employed in various fields such as business, education, ychology, agriculture and economics to mention only a few. When the data eing analyzed is derived from biological sciences and medicine, we use the rm **Biostatistics.**

(b).2.2 Population

ne individual observations from which inferences are made at a particular time e called as a *Population.* In the planning process the investigator should select the population to which the results will be generalized, referred to as the targ population. If for example, we are interested in the weights of all the childr enrolled in a certain elementary school system, our population consists of these weights. Hence, populations are determined or defined by sphere of intere

Populations may be *finite* or *infinite*. If a population consists of fixed numb of values, the population is said to be *finite*. If, on the other hand, a populati consists of an endless succession of values, the population is said to be *infini* Examples: A statistical population may also be birth weights, all women of 15- years of age or only married women, readings of a thermometer, number RBCs in the human body, etc.

26(b).2.3 Sample

Sample may be defined as a part of population that is the representative of t population whose variables are under study. It would be unusual to be able study the entire population of interest, so the researchers select a particular p of the population.

26(b).2.4 Sampling

The method of selecting a sample is known as *Sampling.* Various methods sampling are Simple random sampling, Systemic random sampling, Stratifi random sampling and Cluster random sampling.

26(b).2.5 Variable

If we observe a characteristic, we find that it takes different values on differe persons, places or things. We label this characteristic as a *variable.* We do t for the simple reason that the characteristic of interest is different in differe persons. Some examples include Diastolic BP, Heart rate, the heights of ad males, severity of a disease, and measurement of death and hospital administrati

Variables are further classified into **Quantitative variables** and **Qualitati variables,** which are given below. Individuals or variables that are distinguish by some quality are **Qualitative variables** and those which are distinguish by some measurement are called as **Quantitative variables.**

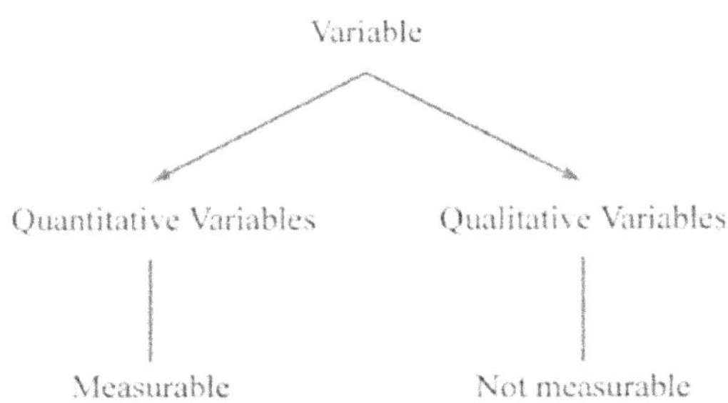

Quantitative variable

A quantitative variable is one that can be measured in the usual sense, which can take numerical values, measurable in nature. Measurements made on quantitative variables convey information regarding amount. Examples: Diastolic BP, Heart rate, the heights of adult males, etc.

Qualitative variable

The variables, which are not measurable in nature, are called qualitative variables. Variables like this can be only categorized. Measurements made on these variables convey information regarding attribute. Examples: Severity of a disease, measurement of death and hospital administration.

Although in the case of qualitative variables, measurement in the usual sense of the word is not achieved. We can count the number of places or things belonging to various categories. These counts or frequencies as they are called are the numbers that we manipulate when our analysis involves qualitative variables.

26(b).2.6 Observation

The individual data points in a variable are called observations.

26(b).2.7 Constant

Quantities that do not vary such as p=3.141, e=2.718. They do not require statistical study. In biostatistics, mean, standard deviation, standard error and correlation coefficient of a particular sample are said to be the constants.

26(b).2.8 Parameter

It is a summary value that describes the population such as mean, standard deviation and proportion.

26(b).2.9 Statistic

It is a summary value that describes the sample such as its mean, standard error and proportion. This value is calculated from the sample and is often applied to the population. Some common notations are given below:

Summary value	Sample Statistic	Population Parameter
Mean	$\overline{x}$	μ
Standard Deviation	s	σ
Proportion	p	P
Variance	s^2	σ^2

26(b).3 Sampling Unit

Each member of a sample is called as a Sampling unit.

26(b).3.1 Classification of data (or) Scope of Measurement

It is important to know the scale of measurement for treatment and response variables because it influences sample size calculation, the choice of appropriate descriptive, relational and comparative statistical techniques. Scale of measurement can be divided into three categories:

(a) Nominal or Categorical

(b) Ordinal or rank order

(c) Continuous (Interval or Ratio)

Nominal Scale (Categorical)

Using a nominal scale of measurement, a variable is measured by counting sample members in a category. Gender is measured on a nominal **scale** - Male or Female. In the early days of computing, numerical values were assigned to categories. Thus, sample mean could be computed for arbitrarily assigned values. Examples: Martial status, diagnostic category and clinical service.

Ordinal Scale (Rank order)

For variables measured on an *ordinal scale*, members of the sample can be ranked in order of magnitude from lowest to highest or from highest to lowest. Results of laboratory tests with negative, 1+, 2+, 3+, and 4+ are examples. Assessments such as mild, moderate or severe and psychological scales with answers such as "Disagree strongly", "Disagree", "No opinion", "agree" or "agree strongly" are ordinal in nature.

Continuous (Interval or Ratio)

Measurements on a continuous scale take one of two forms: interval or ratio. These are purely numerical in nature. If a variable is measured on a continuous scale, the investigator should record the numerical value. Examples include height, weight, elapsed time, change in temperature, duration of illness, etc.

The values of a continuous variable can be reclassified subsequently as nominal or ordinal, if desired. For example, age recorded on subjects can be changed to nominal as, one category may be those subjects under 21years of age, while the other consists of those subjects 21years of age or older. Same variable age can be converted to ordinal as assign group 1 to those subjects, who are less than 18years of age, group 2 to those subjects between 18 and 65 years of age and group 3 to subjects over 65 years of age.

The Classification of data can be easily understood from the following figure.

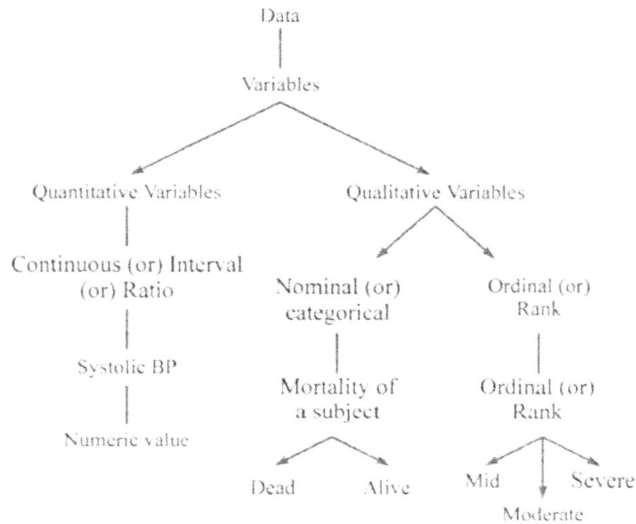

26(b).3.2 Descriptive Statistics

Descriptive statistics are commonly used to present the data baseline characteristics of a population enrolled in a clinical trial. This includes tabulation of results, calculation of averages, calculation of min/max and standard deviations. It mainly includes Central tendency, Dispersion, Skew ness and kurtosis.

Thus, descriptive statistics helps

(a) To get a single value that describes the characteristics of the entire group

(b) How other values are dispersed around the central value?

(c) What is the degree of scatter?

(d) What is the shape of the data, is it normal?

(e) To determine the reliability of an average

(f) To serve as a basis for the control of the variability

(g) To compare two or more series with regard to their variation

(h) To facilitate the use of other statistical measures.

The human mind is incapable of remembering the entire mass of unwieldy data. Therefore, the data has to be summarized into a few summary measures, which can reveal their basic features. One of the most important objectives of this statistical analysis is to get a single value that describes the characteristics of the entire mass of data. It is found in any data, values of the variable tend to congregate around a central value of the data. This tendency of the data is known as *Central tendency.*

The properties that are followed by a central tendency are:

(a) It should be used to understand

(b) It should be simple to compute

(c) It should be based on all the items

(d) It should not be unduly affected by extreme values

(e) It should be rigidly defined

(f) It should be capable of further algebraic treatment

(g) It should have sampling stability

The main three measures of central tendency are – Mean, Median and Mode.

Mean:

This is also known as arithmetic mean. It is obtained by adding all the observations in a variable and by dividing this total by the number of observations. Symbolically,

$$Mean = \frac{x_1 + x_2 + x_3 + \ldots x_n}{n} = \frac{\sum x_i}{n}, where, i = 1, 2 \ldots\ldots n.$$

Where, x = observations in a variable

n = number of observations

Example 1: Erythrocyte sedimentation rates (ESRs) of 7 subjects are, 7, 5, 3, 4, 6, 4 and 5.

The *Mean* is to be given by,

$$Mean = \frac{7 + 5 + 3 + 4 + 6 + 4 + 5}{7} = \frac{34}{7} = 4.86$$

Though mean follows almost all properties of the good measure of central tendency, it has got some limitations.

Median

Median is the middle observation, when all the observations of a variable are arranged either in ascending or in descending order. In case of even number of observations, it is the mean of two mid values. *Median* is a better indicator of central value when one or more of the lowest or highest observations are wide apart or not so evenly distributed. Example1: In the same example as above, ESR values of subjects, after arranging them in ascending order,

3, 4, 4, 5, 5, 6, 7

From the above data, the mid value is 5. So, median = 5.

xample : 4, 6, 8, 10, 12, 14, 32 are ages of some subjects.

From, the above data

$$Mean = \frac{4+6+8+10+12+14+32}{7} = \frac{86}{7} = 12.3$$

The data is ascending order only and mid value is 10.

So, *Median* = 10.

In this case, mean values give a distorted result as one observation 32 is very rge, so the mean as a measure of central tendency should not be considered propriate. To have a better idea of average, one should ignore unduly high servation such as 32 in the above case. If we remove 32, then the mean will come,

$$Mean = \frac{4+6+8+10+12+14}{6} = \frac{54}{6} = 9$$

Therefore *Mean* = 9 which is much closer to *Median* = 10.

Another good example is the duration of stay in a hospital in general or in a ecific disease ward. In such cases, Median is a better indicator because the ay may be unduly long in some cases.

ode

ode is the most frequently occurring observation in a data. *Mode* is rarely ed in medical statistics.

xample: The Systolic BP of 5 subjects is 110, 120, 120, 112, and 115.

Here, *Mode* = 120.

ispersion

may happen such that in several series, the average may be the same but the riables may highly differ in magnitude, and, therefore, the central tendency ay not be the true representative.

For example, consider the following two series of ages of 5 subjects.

Series A: 18, 18, 18, 18, and 18.

Series B: 12, 17, 18, 18, and 25.

From the above data, for both the series A and B *Mean, Median* and *Mode* are equal to 18. But, the data is not uniform in series B compared to series A.

Dispersion is the measure of variation in the variable. It measures the extent which there are differences between individual observations and central value. e averages cannot adequately describe a set of observations, unless all servations are the same. In two or more samples of data, the central tendency y be the same but there can be wide disparity in the formation of data.

It is clear that dispersion measures the extent to which central values var from each other. This is explained from the following example.

Series A	Series B
100	1
105	489
102	2
103	3
90	5
$\bar{x} = 100$	$\bar{x} = 100$

Since, *Mean* is the same in all the series, one is likely to conclude that thes series all are alike in nature. But a close examination will reveal that the da differ widely from one another. This can be shown in the normal curves belo

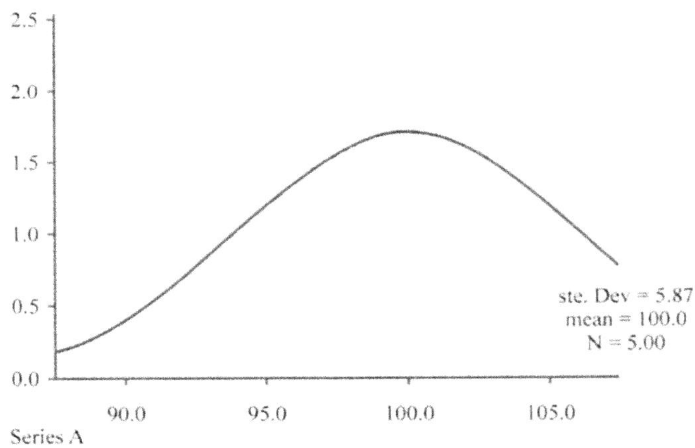

From the above diagram, we can observe that Series A is more uniform more homogenous or less dispersed compared to Series B.

The following are the important measures for studying variation or dispersio

1. Range
2. Quartile Deviation
3. Mean Deviation
4. Variance

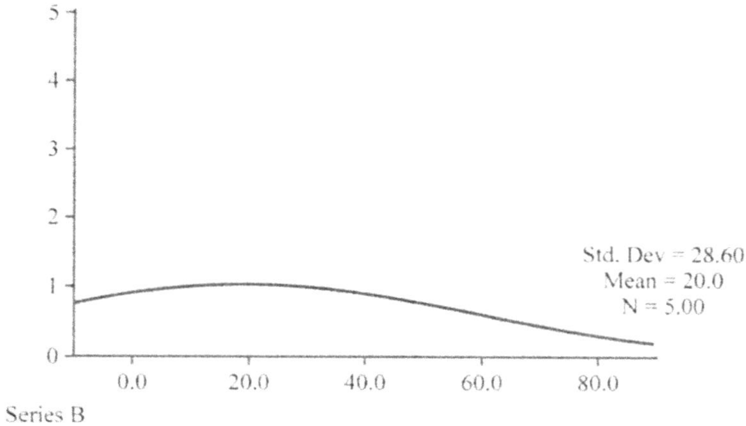

Of these, Range and Quartile Deviation are positional measures as they depend on the values at a particular position in the data. Mean Deviation and Variance are called calculation measures of dispersion because all the values of the data are employed in their calculation.

Range

Range is defined as the difference between the lowest and highest values of the variables in the data.

Symbolically,

$$Range = L\text{-}S,$$

where L is the Largest value and S is the Smallest value

Example: Ages of subjects in years are given below.

20, 18, 22, 27, 19, 17.

Range = Largest-Smallest = 27 – 17 = 10

Generally, observations within a particular *Range* are considered normal and those falling outside the range are called abnormal. *Range* is not a satisfactory measure as it is based on only two extreme values, ignoring the remaining data within the extremes.

Quartile Deviation (QD)

When Q1 and Q3 are the first and third quartiles of distribution respectively, then the *Quartile deviation* is given by

$$QD = \frac{1}{2}(Q_3 - Q_1)$$

Quartile deviation is definitely a better measure than Range as it makes use of 50% of the data. But since it ignores the remaining 50% of the data, it cannot be regarded as a reliable measure.

Mean Deviation (MD)

The *Mean deviation* is also known as *Average deviation.* It is calculated either from the mean or the median. Mean is more commonly used in calculating the value of the mean deviation. It is obtained by calculating the absolute difference of each observation from the mean or median and then averaging these deviations by taking their mean.

Symbolically,

Sample Mean deviation (MD) = $\frac{\sum|x-\bar{x}|}{n}$,

where x = observations, $\bar{x}$ = Sample mean and n = number of observations

The reason for taking mean in the calculation of MD is that the sum of deviations taken from the mean is less than the sum of deviations from any central tendency.

Though MD satisfies almost all the properties of *Dispersion,* there are some disadvantages. They are:

1. By considering absolute values, it is ignoring the algebraic signs of deviations.
2. This method may not give accurate results when the MD is taken from *Median*, which is not a satisfactory measure when the degree of variability is very high.

To overcome all the above disadvantages, another measure of dispersion was introduced, called *Variance*.

Variance: It is also called Mean Square Deviation. To calculate the *Variance* the deviations of the variables from the mean are squared and then added. The number of observations to get the variance of the sample, which is denoted by s?, divides the sum of squares of deviations.

To put it simply,

Sample variance, $s^2 = \frac{\sum(x-\bar{x})^2}{n-1}$ where x = observations,

$\bar{x}$ = Sample mean and n = number of observations

This measure has an advantage over MD, because the sum of squares of leviations is always positive.

The reason for taking n-1 as the denominator in calculation of variance is hat, when size of the sample is small, the difference will be relatively high. Iowever, in the samples of large size, the Variance will show little difference. To get a good estimate of Variance or to make it unbiased towards sample size, t has to be divided by n-1.

The disadvantage of Variance is that it is not included in the Absolute measures of dispersion as it is not expressed in the same units in which the original data is measured.

Example: As it relates to squared deviations, the variance of the yield of the plant measured in grams should have been expressed in squared grams. This practical difficulty can be over come by taking the square root of Variance, which is called Standard deviation.

Standard Deviation (SD)

Standard deviation is defined as the square root of variance. It is the most commonly used absolute measure of dispersion.

Symbolically,

Sample SD, $s = \sqrt{\frac{\sum(x - \bar{x})^2}{n - 1}}$ where x = observations,

$\bar{x}$ = Sample mean and n = number of observations

SD is expressed in the same units in which the original data is calculated. As t measures the *dispersion* or variability of the data, larger the SD, the grater is he variability. On the other hand, there will be homogeneity in a series when the SD is small. To put it simply, SD is directly proportional to variability or dispersion and inversely proportional to homogeneity.

The SD is an absolute measure of *dispersion*. But, the SD of height of plants cannot be compared with the SD of the weight of plants because they are expressed in different units, i.e. height (cms) and weight (gms). Therefore, the SD must be converted into a relative measure of *dispersion* for the purpose of comparison. The relative measure of dispersion is known as Coefficient of Variation.

Coefficient of variation

Coefficient of variation is of great practical significance and is the best measure of comparing the variability or consistency of two samples. It is expressed as he percentage of the ratio of sample SD (s) to mean ().

Symbolically,

$$CV = \frac{s}{\bar{x}} * 100$$

Coefficient of variation works similar to SD. The smaller the CV, the greater is its consistency. On the other hand, if the CV is greater the sample is said to be more variable or less homogenous.

Example: For two series of data, the mean ($\bar{x}$) and SD (s) are given below.

Series A: $\bar{x}$ = 34.6 and s = 14

Series B: $\bar{x}$ =27.1 and s = 8

From the above data,

$$CV (A) = \frac{34.6}{14} = 40.46\ \% \text{ and } CV (B) = \frac{27.1}{8} = 29.52\%$$

From the above results we can conclude that Series B is more homogenous compared to Series A.

Skewness and Kurtosis

Measures of Central tendency and dispersion do not reveal the main characteristics of the measurement of shape. The measurement of shape involves two characteristics: Skewness, which refers to the lack of symmetry of the data, and Kurtosis, which refers to the extent to which data is peaked.

Skewness

Skewness is defined as the lack of symmetry in the data. When the two samples of the data have the same mean and SD, they may differ in the overall performance mainly due to symmetry. In symmetrical data all the values of Central tendency i.e. Mean, Median and Mode are equal. The more the mean moves away from the mode, the large the asymmetry or skewness. If the data is not symmetrical it is said to be asymmetrical or skewed. Therefore, the measures of central tendency and dispersion should always be supplemented by the measures of skewness to distinguish between different types of data. In a skewed data, the mean, median and mode fall at different places.

There are three types of skewness namely Symmetric, Positive and Negative skewness. They are explained below.

Symmetric data

Data is said to be symmetrical, when all the central tendency values are equal i.e. Mean = Median = Mode. In this data, we will get a bell-shaped curve.

Example: Consider the ages of 5 subjects 18, 16, 18, 20, 18.

From the above data,

$$Mean = \frac{18+16+18+20+18}{5} = \frac{90}{5} = 18$$

After arranging the data in ascending order 16,18,18,18,20; the mid value, ledian=18 and the repeating value, Mode = 18.

This can be seen from the following curve not clear

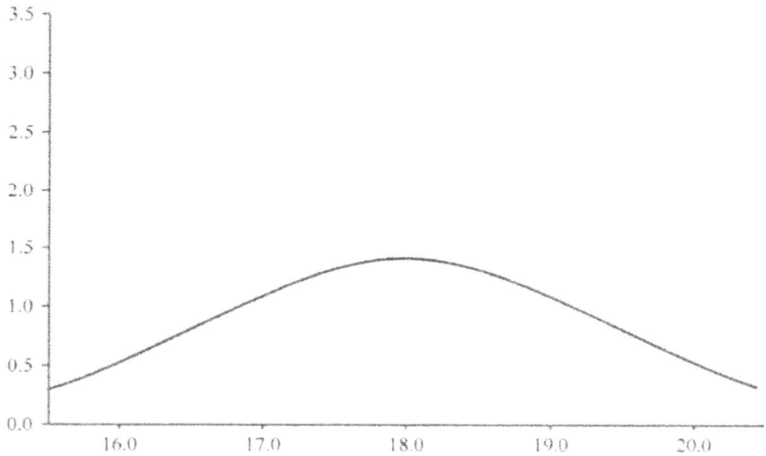

Symmetric data where Mean = Median = Mode = 18

ositively skewed data

ata is said to be *positively skewed,* when Mean>Median>Mode. In this, the ırve is stretched more towards right.

xample: Consider the Systolic BP of 5 subjects 110, 120, 130, 110, 140

From the above data, *Mean* = 122, *Median* = 120 and *Mode* = 110.

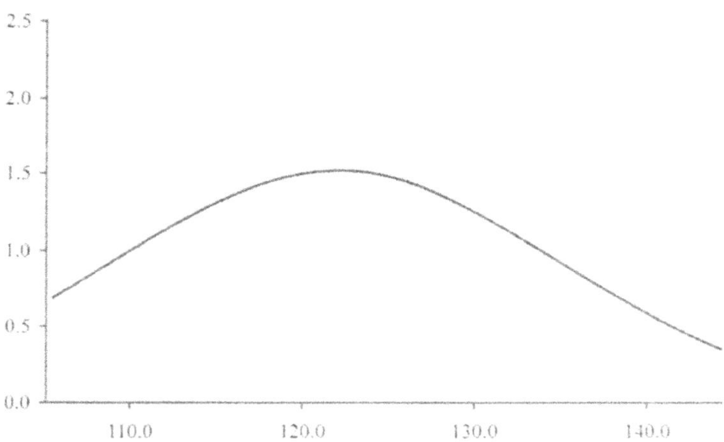

Postively Skewed where Mean > Median > Mode

Negatively Skewed data

Data is said to be negatively skewed, when mean<Median<Mode. In this, th curve is stretched more towards left.

Example: Consider the ages of 9 subjects 11, 14, 16,17,10,19,20,17,20

From the above data, *Mean =16, Median = 17 and Mode = 17 or 20*

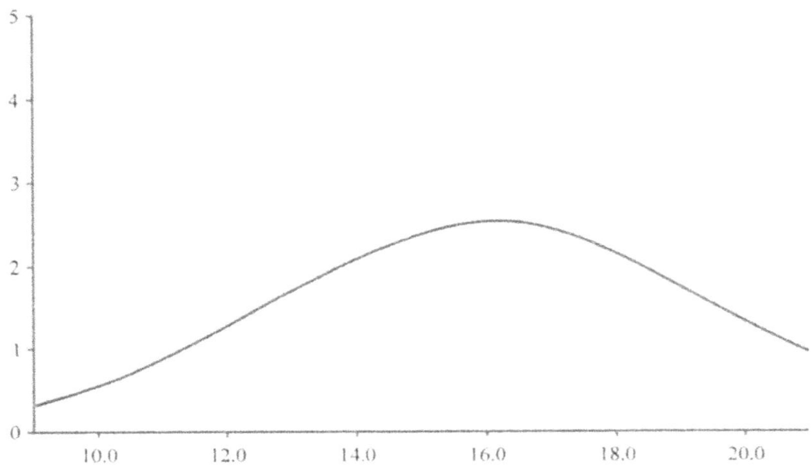

Nagatively skewed where Mean < Median < Mode

Kurtosis

Kurtosis is the other measure of shape, which tells us whether the data is flat peaked. There are three types of kurtosis namely: Mesokurtic, Leptokurtic, a Platykurtic.

Mesokurtic

A normal curve is a smooth, continuous, perfect symmetrical bell-shaped curv which is called Mesokurtic.

Leptokurtic

If a curve is more peaked than normal curve, it is called Leptokurtic.

Platykurtic

When a curve is more flat than normal curve it is called Platykurtic.

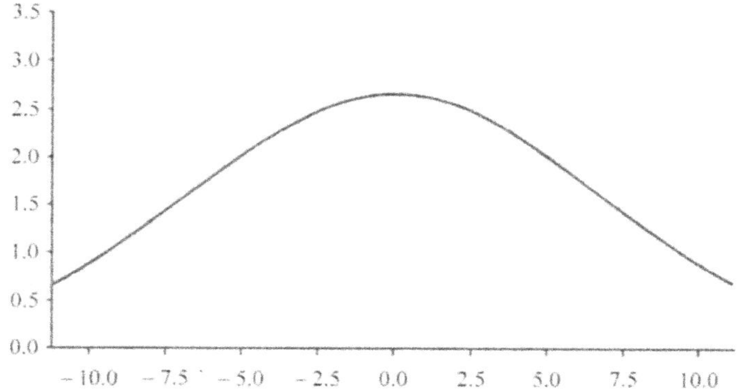

Mesokurtic

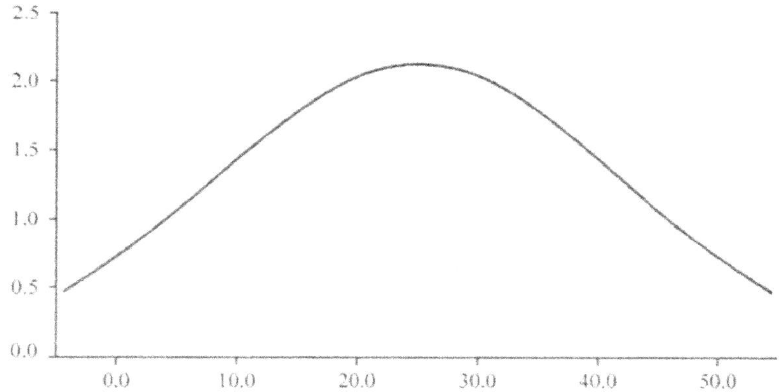

Leptokurtic

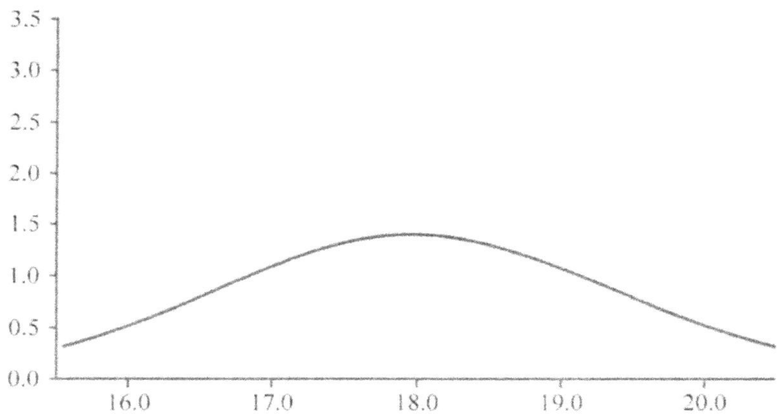

Platykurtic

Various Statistical Tests Used

Test Statistics

Test statistics is a quantity calculated from our sample of data. Its value is usec to decide whether or not the null hypothesis should be rejected in our hypothesi test. The choice of a test statistic will depend on the assumed probability mode and the hypothesis under question.

Critical Value(s)

The critical value(s) for a hypothesis test is a threshold to which the value of th test statistic in a sample is compared to determine whether or not the nul hypothesis is rejected. The critical value for any hypothesis test depends on th significance level at which the test is carried out, and whether the test is one sided or two-sided.

Critical Region

The critical region CR, or rejection region RR, is a set of values of the tes statistics for which the null hypothesis is rejected in a hypothesis test. That is the sample space for the test statistics is partitioned into two regions; one regio (the critical region) will lead us to reject the null hypothesis H0, the other wi not. So, if the observed value of the test statistics is a member of the critica region, we conclude "Reject H0"; if it is not a member of the critical region the we conclude "Do not reject H0".

Significance Level

The significance level of a statistical hypothesis test is a fixed probability c erroneously rejecting the null hypothesis H0, if it is in fact true.

It is the probability of a *type I error* and is set by the investigator in relatio to the consequences of such an error. That is, we want to make the significanc level as small as possible in order to protect the null hypothesis and to preven as far as possible, the investigator from inadvertently making false claims.

The significance level is usually denoted by

Significance Level = p (type I error)

Usually, the significance level is chosen to be 0.05 (or equivalently, 5%).

p-Value

The probability value (p-value) of a statistical hypothesis test is the probabili of getting a value of the test statistic as extreme as or more extreme than th observed by chance alone, if the null hypothesis H0, is true.

It is the probability of wrongly rejecting the null hypothesis if it is in fact tru

It is equal to the significance level of the test for which we would only just eject the null hypothesis. The p-value is compared with the actual significance evel of our test and, if it is smaller, the result is significant. That is, if the null ypothesis were to be rejected at the 5% significance level, this would be reported s "$p < 0.05$".

Small p-values suggest that the null hypothesis is unlikely to be true. The maller it is, the more convincing is the rejection of the null hypothesis. It indicates he strength of evidence for say, rejecting the null hypothesis H0, rather than imply concluding "Reject H0' or "Do not reject H0".

ower

he power of a statistical hypothesis test measures the test's ability to reject the ull hypothesis when it is actually false - that is, to make a correct decision.

In other words, the power of a hypothesis test is the probability of not ommitting a ***type II error.*** It is calculated by subtracting the probability of a ype II error from 1, usually expressed as:

$$\text{Power} = 1 - P(\text{type II error}) = (1 - \beta)$$

The maximum power a test can have is 1, the minimum is 0. Ideally we want test to have high power, close to 1.

ne-sided Test

ne-sided test is a test of statistical hypothesis in which we can reject the null ypothesis, H0, where the values are located entirely in one tail of the probability istribution. In other words, the critical region for a one-sided test is the set of alues less than the critical value of the test, or the set of values greater than the ritical value of the test. A one-sided test is also referred to as a one-tailed test f significance.

The choice between a one-sided and a two-sided test is determined by the urpose of the investigation or prior reasons for using a one-sided test.

xample

uppose we wanted to test a manufacturers claim that there are, on average, 50 atches in a box. We could set up the following hypothesis

H0: average = 50, against H1: average < 50 or H1: average > 50

Either of these two alternative hypothesis would lead to a one-sided test. resumably, we would want to test the null hypothesis against the first alternative ypothesis since it would be useful to know if there is likelihood of having less an 50 matches, on average, in a box (no one would complain if they get the rrect number of matches in a box or more).

Yet another alternative hypothesis could be tested against the same null, leading this time to a two-sided test:

H0: average = 50, against H1: average not equal to 50

Here, nothing specific can be said about the average number of matches in a box; only that, if we could reject the null hypothesis in our test, we would know that the average number of matches in a box is likely to be less than or greater than 50.

Two-Sided Test

A two-sided test is a test of statistical hypothesis in which we can reject the null hypothesis, H0, where the values are located in both tails of the probability distribution.

In other words, the critical region for a two-sided test is the set of values less than a first critical value of the test and the set of values greater than a second critical value of the test.

A two-sided test is also referred to as a two-tailed test of significance.

The choice between a one-sided test and a two-sided test is determined by the purpose of the investigation or prior reasons for using a one-sided test.

Example

Suppose we wanted to test a manufacturers claim that there are, on average, 50 matches in a box. We could set up the following hypothesis

H0:average $\mu = 50$, against H1:average < 50 or H1: average > 50

Either of these two alternative hypothesis would lead to a one-sided test. Presumably, we would want to test the null hypothesis against the first alternative hypothesis since it would be useful to know if there is a likelihood of having less than 50 matches, on average, in a box (no one would complain if they get the correct number of matches in a box or more).

Yet another alternative hypothesis could be tested against the same null, leading this time to a two-sided test:

H0:average = 50, against H1: average not equal to 50

Here, nothing specific can be said about the average number of matches in a box; only that, if we could reject the null hypothesis in our test, we would know that the average number of matches in a box is likely to be less than or greater than 50.

One Sample t-test

A one sample t-test is a hypothesis test for answering questions about the mean where the data are a random sample of independent observations from an underlying normal distribution $N(\mu,\sigma^2)$, where σ^2 is unknown.

The null hypothesis for the one sample t-test is:

H0: $\mu = \mu 0$, where $\mu 0$ is known.

That is, the sample has been drawn from a population of a given mean and unknown variance (which therefore has to be estimated from the sample).

This null hypothesis, H0 is tested against one of the following alternative hypothesis, depending on the question posed:

H1: μ is not equal to μ

H1: $\mu > \mu$

H1: $\mu < \mu$

Two Sample t-test

A two sample t-test is a hypothesis test for answering questions about the mean where the data are collected from two random samples of independent observations, each from an underlying normal distribution:

$$N\left(\mu_{i,}\ \sigma_i^2\right) \text{ where } i = 1, 2$$

When carrying out a two sample t-test, it is usual to assume that the variances for the two populations are equal, i.e.

$$\sigma_1^{\ 2} = \sigma_2^{\ 2} = \sigma^2$$

The null hypothesis for the two sample t-test is:

H0: $\mu 1 = \mu 2$

That is, the two samples have both been drawn from the same population. This null hypothesis is tested against one of the following alternative hypothesis, depending on the question posed.

H1: $\mu 1$ is not equal to $\mu 2$

H1: $\mu 1 > \mu 2$

H1: $\mu 1 < \mu 2$

Non-parametric Tests

Non-parametric tests are often used in place of their parametric counterparts when certain assumptions about the underlying population are questionable. For example, when comparing two independent samples, the Wilcoxon Mann-Whitney test does not assume that the difference between the samples is normally distributed whereas its parametric counterpart, the two sample t-test does. Non-parametric tests may be, and often are, more powerful in detecting population differences when certain assumptions are not satisfied.

All tests involving ranked data, i.e. data that can be put in order, ar nonparametric.

Mann-Whitney U Test

The Mann-Whitney Test U is one of the most powerful nonparametric tests f comparing two populations. It is used to test the null hypothesis, that is, tw populations have identical distribution functions against the alternative hypothesi that the two distribution functions differ only with respect to location (median), i at all.

The Mann-Whitney U test does not require the assumption that the difference between the two samples are normally distributed.

In many applications, the Mann-Whitney Test U is used in place of the tw sample t-test when the normality assumption is questionable. This test can als be applied when the observations in a sample of data are ranks, that is, ordin data rather than direct measurements.

Wilcoxon Signed Rank Test

The Wilcoxon Signed Rank test is designed to test a hypothesis about the locatio (median) of a population distribution. It often involves the use of matched pair for example, before and after data, in which case it tests for a median differenc of zero. The Wilcoxon Signed Rank test does not require the assumption that th population is normally distributed. In many applications, this test is used in plac of the one sample t-test when the normality assumption is questionable. It is more powerful alternative to the sign test, but does assume that the populatic probability distribution is symmetric. This test can also be applied when th observations in a sample of data are ranks, that is, ordinal data rather than dire measurements.

Sign Test

The sign test is designed to test a hypothesis about the location of a populatic distribution. It is most often used to test the hypothesis about a population media and often involves the use of matched pairs, for example, before and after dat in which case it tests for a median difference of zero.

The Sign test does not require the assumption that the population is normal distributed.

In many applications, this test is used in place of the one sample t-test wh the normality assumption is questionable. It is a less powerful alternative to t Wilcoxon signed rank test, but does not assume that the population probabili distribution is symmetric.

This test can also be applied when the observations in a sample of data a ranks, that is, ordinal data rather than direct measurements.

Kruskal-Wallis Test

The Kruskal-Wallis test is a nonparametric test used to compare three or more samples. It is used to test the null hypothesis that all populations have identical distribution functions against the alternative hypothesis that at least two of the samples differ only with respect to location (median), if at all.

It is the analogue to the F-test used in analysis of variance. While analysis of variance tests depend on the assumption that all populations under comparison are normally distributed, the Kruskal-Wallis test places no such restriction on the comparison.

Analysis of Variance - Anova

The t test is generalized to more than two groups by means of a technique termed as the analysis of variance. In this method there are both between and within groups of degree of freedom.

One way ANOVA

One way ANOVA is known as one way analysis of variance, a powerful tool for testing the significance. It is an extension to three or more samples of the t-test procedure (useful to find the significance of two independent samples). The hypothesis of interest is equality of the means of different samples. The term ANOVA was introduced by Prof. R. A. Fisher in 1920 to deal with a problem agronomical in nature.

Assumptions: Although, ANOVA is a robust method (much difference wont be there, even the data violates one or more of the below mentioned assumptions) in testing the significance, the following assumptions are made in order to test the hypothesis,

1. The observations are independent
2. Parent population is normal
3. Various treatments and environmental effects are additive in nature.

Example: In an experiment, 4 dogs were treated with estrogen, another 4 dogs were treated with progesterone, 5 dogs were treated with estrogen plus progesterone and 5 dogs untreated served as control. In all animals serum progesterone levels were determined.

We assume that the four sets of data constitute independent random sample from four populations that are similar except for the treatment received.

We frame hypothesis as

H_0: The effect of all the four treatment groups is equal, against

H_1: The effect all the four treatment groups is not equal.

Serum progesterone levels in various groups

	Untreated control	Estrogen group	Progesterone Group	Estrogen+ Progesterone (EP)	
	117	440	605	2664	
	124	264	626	2078	
	40	221	385	3584	
	88	136	475	1540	
	40	-	-	1840	Row total
Column Total	409	1061	2091	11706	15267
Mean	81.80	265.25	522.75	2341.20	848.17

Here, the test statistic is given by,

VR = MSA / MSW,

where, VR = Variance Ratio

MSA = Mean Square Among the groups (between groups)

MSW = Within group Mean Square

F = Statistically significance level, it is given from F table.

In the present example, we wish to see the significance at the level á = 0.0 (95%). Thus from the table a total 4 groups are there, so 4 -1 = 3 numerat degrees of freedom and total observations 18 - 4 = 14 denominator degrees freedom. At this degree of freedom, F value from F table = 3.34.

If the value of variance ratio is more than 3.34, the result is considered to b significant.

Now, let us follow the steps to calculate the variance ratio.

Step 1

Find out SST, the Total sum of squares

1. Square all individual observations and add – I
2. Square the total of 18 observations and divide it by the number observations – II
3. Subtract the value of II from I

From the example,

$(117)^2 + (124)^2 + (40)^2 + (88)^2 + (40)^2 + (440)^2 + (264)^2 + (221)^2 + (136)^2 + (605)^2 + (626)^2 + (385)^2 + (475)^2 + (2664)^2 + (2078)^2 + (3584)^2 + (1540)^2 + (1840)^2 = 31519629 - I$

117 + 124 + 40 + 88 + 40 + 440 6 264 + 221 + 136 + 605 + 626 + 385 + 475 + 2664 + 2078 + 3584 + 475 + 2664 + 2078 + 3584 + 1540 + 1840 = 15267

$$(15267)^2/18 = 12948960.5 - II$$

So, SST = 31519629 – 12948960.5 = 18570668.5

Step 2

Calculate the SSA, among group sum squares or sum of squares between the groups

1. Square the column totals and divide it with the respective number of observations and add these values - I
2. Take the row total of column totals, square it and divide it with the number of observations - II
3. Subtract II from I

From the example,

$$(409)^2 / 5 + (1062)^2 / 4 + (2091)^2 / 4 + (11706)^2 / 5 - (15267)^2 / 18$$

$$= 28814043.9 - 12948960.5 = 15865083.4$$

So, SSA = 15865083.4

Step 3

Calculate SSW, within group sum of squares.

This sum of squares is obtained by subtracting among group sum of squares, SSA from total sum of squares, SST.

Now, SSW = SST – SSA

= 18570668.5 – 15865083.4

= 2705585.1

Now insert all these values in the below table.

Source	Sum of squares (1)	Degrees of freedom (2)	Mean sum of squares (1) ÷ (2)	Variance ratio
Among samples	15865083.4	3	5288361.133	27.3645
Within sample	2705585.1	14	193256.0786	

Variance ratio (VR) is obtained by dividing among group mean sum of squares with within group mean sum of squares.

VR = among group mean sum of squares / within group mean sum of squares

= 5288361.133 / 193256.0786

= 27.3645

Now, the F-value at 0.05 significant level from the F table is 3.34. In this study, since our variance ratio is 27.3645, which is more than 3.34, we do not accept the null hypothesis, H_0, thus concluding there is a significant difference between the four groups.

Suggested Reading

1. Hill AB. Principles of medical statistics. 9ed, London: The Lancet 1971
2. Anello C. Statistical issues in Comparative Clinical Clinical Trials. Drug Inf. 1983; 17:55
3. Armitage P. Statistical methods in medical research. New York; Willy and Sons 1973.
4. Buncher CR and Tsay JY. Statistics in Pharmaceutical Industry. NewYork: Marcel Dekkar 1981.

Glossary

Adverse Drug Reaction (ADR)

In the pre-approval clinical experience with a new medicinal product or its new usages, particularly as the therapeutic dose(s) may not be established: all noxious and unintended responses to a medicinal product related to any dose should be considered adverse drug reactions. The phrase responses to a medicinal product means that a causal relationship between a medicinal product and an adverse event is at least a reasonable possibility, i.e. the relationship cannot be ruled out. Regarding marketed medicinal products: a response to a drug which is noxious and unintended and which occurs at doses normally used in man for prophylaxis, diagnosis, or therapy of diseases or for modification of physiological function (see the ICH Guideline for Clinical Safety Data Management: Definitions and Standards for Expedited Reporting).

Adverse Event (AE)

Any untoward medical occurrence in a patient or clinical investigation subject administered a pharmaceutical product and which does not necessarily have a causal relationship with this treatment. An adverse event (AE) can therefore be any unfavourable and unintended sign (including an abnormal laboratory finding), symptom, or disease temporally associated with the use of a medicinal (investigational) product, whether or not related to the medicinal (investigational) product (see the ICH Guideline for Clinical Safety Data Management: Definitions and Standards for Expedited Reporting).

Amendment (to the Protocol)

See Protocol Amendment.

Applicable Regulatory Requirement(s)

Any law(s) and regulation(s) addressing the conduct of clinical trials of investigational products.

Approval (in relation to Institutional Review Boards)

The affirmative decision of the IRB that the clinical trial has been reviewed and may be conducted at the institution site within the constraints set forth by the IRB, the institution, Good Clinical Practice (GCP), and the applicable regulatory requirements.

Audit

A systematic and independent examination of trial related activities and documents to determine whether the evaluated trial related activities were conducted, and the data were recorded, analyzed and accurately reported according to the protocol, sponsor's standard operating procedures (SOPs), Good Clinical Practice (GCP), and the applicable regulatory requirement(s).

Audit Certificate

A declaration of confirmation by the auditor that an audit has taken place.

Audit Report

A written evaluation by the sponsor's auditor of the results of the audit.

Audit Trail

Documentation that allows reconstruction of the course of events.

Blinding/Masking

A procedure in which one or more parties to the trial are kept unaware of the treatment assignment(s). Single-blinding usually refers to the subject(s) being unaware, and double-blinding usually refers to the subject(s), investigator(s), monitor, and, in some cases, data analyst(s) being unaware of the treatment assignment(s).

Case Report Form (CRF)

A printed, optical, or electronic document designed to record all of the protocol required information to be reported to the sponsor on each trial subject.

Clinical Trial/Study

Any investigation in human subjects intended to discover or verify the clinical, pharmacological and/or other pharmacodynamic effects of an investigational product(s), and/or to identify any adverse reactions to an investigational product(s), and/or to study absorption, distribution, metabolism, and excretion of an investigational product(s) with the object of ascertaining its safety and/or efficacy. The terms clinical trial and clinical study are synonymous.

Clinical Trial/Study Report

A written description of a trial/study of any therapeutic, prophylactic, or diagnostic agent conducted in human subjects, in which the clinical and statistical description, presentations, and analyses are fully integrated into a single report (see the ICH Guideline for Structure and Content of Clinical Study Reports).

Comparator (Product)

An investigational or marketed product (i.e., active control), or placebo, used as a reference in a clinical trial.

Compliance (in relation to trials)

Adherence to all the trial-related requirements, Good Clinical Practice (GCP) requirements, and the applicable regulatory requirements.

Confidentiality

Prevention of disclosure, to other than authorized individuals, of a sponsor's proprietary information or of a subject's identity.

Contract

A written, dated, and signed agreement between two or more involved parties that sets out any arrangements on delegation and distribution of tasks and obligations and, if appropriate, on financial matters. The protocol may serve as the basis of a contract.

Coordinating Committee

A committee that a sponsor may organize to coordinate the conduct of a multicentre trial.

Coordinating Investigator

An investigator assigned the responsibility for the coordination of investigators at different centres participating in a multicentre trial.

Contract Research Organization (CRO)

A person or an organization (commercial, academic, or other) contracted by the sponsor to perform one or more of a sponsor's trial-related duties and functions.

Direct Access

Permission to examine, analyzes, verify, and reproduce any records and report that are important to evaluation of a clinical trial. Any party (e.g., domestic an foreign regulatory authorities, sponsor's monitors and auditors) with direct acces should take all reasonable precautions within the constraints of the applicabl regulatory requirement(s) to maintain the confidentiality of subjects' identitie and sponsor's proprietary information.

Documentation

All records, in any form (including, but not limited to, written, electronic, magneti and optical records, and scans, x-rays, and electrocardiograms) that describe c record the methods, conduct, and/or results of a trial, the factors affecting trial, and the actions taken.

Essential Documents

Documents which individually and collectively permit evaluation of the condu of a study and the quality of the data produced (see 8. Essential Documents fc the Conduct of a Clinical Trial).

Good Clinical Practice (GCP)

A standard for the design, conduct, performance, monitoring, auditing, recordin analyses, and reporting of clinical trials that provides assurance that the da and reported results are credible and accurate, and that the rights, integrity, an confidentiality of trial subjects are protected.

Independent Data-Monitoring Committee (IDMC) (Data and Safe Monitoring Board, Monitoring Committee, Data Monitoring Committe

An independent data-monitoring committee that may be established by th sponsor to assess at intervals the progress of a clinical trial, the safety data, ar the critical efficacy endpoints, and to recommend to the sponsor whether continue, modify, or stop a trial.

Impartial Witness

A person, who is independent of the trial, who cannot be unfairly influenced b people involved with the trial, who attends the informed consent process if th subject or the subject's legally acceptable representative cannot read, and wh reads the informed consent form and any other written information supplied the subject.

Independent Ethics Committee (IEC)

An independent body (a review board or a committee, institutional, region national, or supranational), constituted of medical professionals and non-medic members, whose responsibility it is to ensure the protection of the rights, safety and well-being of human subjects involved in a trial and to provide public assurance of that protection, by, among other things, reviewing and approving / providing favourable opinion on, the trial protocol, the suitability of the investigator(s), facilities, and the methods and material to be used in obtaining and documenting informed consent of the trial subjects. The legal status, composition, function, operations and regulatory requirements pertaining to Independent Ethics Committees may differ among countries, but should allow the Independent Ethics Committee to act in agreement with GCP as described in this guideline.

Informed Consent

A process by which a subject voluntarily confirms his or her willingness to participate in a particular trial, after having been informed of all aspects of the trial that are relevant to the subject's decision to participate. Informed consent is documented by means of a written, signed and dated informed consent form.

Inspection

The act by a regulatory authority(ies) of conducting an official review of documents, facilities, records, and any other resources that are deemed by the authority(ies) to be related to the clinical trial and that may be located at the site of the trial, at the sponsor's and/or contract research organization's (CRO's) facilities, or at other establishments deemed appropriate by the regulatory authority(ies).

Institution (medical)

Any public or private entity or agency or medical or dental facility where clinical trials are conducted.

Institutional Review Board (IRB)

An independent body constituted of medical, scientific, and non-scientific members, whose responsibility is to ensure the protection of the rights, safety and well-being of human subjects involved in a trial by, among other things, reviewing, approving, and providing continuing review of trial protocol and amendments and of the methods and material to be used in obtaining and documenting informed consent of the trial subjects.

Interim Clinical Trial/Study Report

A report of intermediate results and their evaluation based on analyses performed during the course of a trial.

Investigational Product

A pharmaceutical form of an active ingredient or placebo being tested or used as a reference in a clinical trial, including a product with a marketing authorization when used or assembled (formulated or packaged) in a way different from the approved form, or when used for an unapproved indication, or when used to gain further information about an approved use.

Investigator

A person responsible for the conduct of the clinical trial at a trial site. If a trial is conducted by a team of individuals at a trial site, the investigator is the responsible leader of the team and may be called the principal investigator. See also Subinvestigator.

Investigator / Institution

An expression meaning "the investigator and/or institution, where required by the applicable regulatory requirements".

Investigator's Brochure

A complication of the clinical and nonclinical data on the investigational product(s) which is relevant to the study of the investigational product(s) in human subjects (see Investigator's Brochure).

Legally Acceptable Representative

An individual or juridical or other body authorized under applicable law to consent on behalf of a prospective subject, to the subject's participation in the clinical trial.

Monitoring

The act of overseeing the progress of a clinical trial, and of ensuring that it is conducted, recorded, and reported in accordance with the protocol, Standard Operating Procedures (SOPs), Good Clinical Practice (GCP), and the applicable regulatory requirement(s).

Monitoring Report

A written report from the monitor to the sponsor after each site visit and/or other trial related communication according to the sponsor's SOPs.

Multicentre Trial

A clinical trial conducted according to a single protocol but at more than one site and therefore, carried out by more than one investigator.

Nonclinical Study

Biomedical studies not performed on human subjects.

Opinion (in relation to Independent Ethics Committee)

The judgment and/or the advice provided by an Independent Ethics Committee (IEC).

Original Medical Record

See Source Documents.

Protocol

A document that describes the objective(s), design, methodology, statistical considerations, and organization of a trial. The protocol usually also gives the background and rationale for the trial, but these could be provided in other protocol referenced documents. Throughout the ICH GCP Guideline the term protocol refers to protocol and protocol amendments.

Protocol Amendment

A written description of a change(s) to or formal clarification of a protocol.

Quality Assurance (QA)

All those planned and systematic actions that are established to ensure that the trial is performed and the data are generated, documented (recorded), and reported in compliance with Good Clinical Practice (GCP) and the applicable regulatory requirement(s).

Quality Control (QC)

The operational techniques and activities undertaken within the quality assurance system to verify that the requirements for quality of the trial-related activities have been fulfilled.

Randomization

The process of assigning trial subjects to treatment or control groups using an element of chance to determine the assignments in order to reduce bias.

Regulatory Authorities

Bodies having the power to regulate. In the ICH GCP guideline the expression Regulatory Authorities includes the authorities that review submitted clinical data and those that conduct inspections (see 1.29). These bodies are sometimes referred to as competent authorities.

Serious Adverse Event (SAE) or Serious Adverse Drug Reaction (Serious ADR)

Any untoward medical occurrence that atwith any dose:

- Results in death,

- Is life-threatening,
- Requires inpatient hospitalization or prolongation of existing hospitalizatio
- Results in persistent or significant disability/incapacity, or
- Is a congenital anomaly/birth defect (see the ICH Guideline for Clinic Safety Data Management: Definitions and Standards for Expedite Reporting).

Source Data

All information in original records and certified copies of original records c clinical findings, observations, or other activities in a clinical trial necessary fc the reconstruction and evaluation of the trial. Source data are contained in sourc documents (original records or certified copies).

Source Documents

Original documents, data, and records (e.g., hospital records, clinical and offic charts, laboratory notes, memoranda, subjects' diaries or evaluation checklist pharmacy dispensing records, recorded data from automated instruments, copie or transcriptions certified after verification as being accurate copies, microfiche photographic negatives, microfilm or magnetic media, x-rays, subject files, an records kept at the pharmacy, at the laboratories and at medico-technic departments involved in the clinical trial).

Sponsor

An individual, company, institution, or organization which takes responsibility f the initiation, management, and/or financing of a clinical trial.

Sponsor-Investigator

An individual who both initiates and conducts, alone or with others, a clinic trial, and under whose immediate direction the investigational product administered to, dispensed to, or used by a subject. The term does not inclu any person other than an individual (e.g., it does not include a corporation or agency). The obligations of a sponsor-investigator include both those of a spons and those of an investigator.

Standard Operating Procedures (SOPs)

Detailed written instructions to achieve uniformity of the performance of a speci function.

Subinvestigator

Any individual member of the clinical trial team designated and supervised the investigator at a trial site to perform critical trial-related procedures and to make important trial-related decisions (e.g., associates, residents, resear fellows). See also Investigator.

ubject/Trial Subject (Participant)

n individual who participates in a clinical trial, either as a recipient of the nvestigational product(s) or as a control. WHO (2000) also defines a research articipant " An individual who participate in thea biomedical research project, ither as direct recipient of an intervention (e.g., study product or invasive rocedure) as a control, through observation. The individual may be healthy erson who volunteers to participate in the research, or a person with a condition nrelated to research carried out who volunteers to participate, or a person usually a patient) whose condition is relevant to the use of the study product or uestion being investigated". (Note: The Guidelines for GCP in the conduct of rials in Human Participants in SA use the term "participant" to refer to "subject r trial subject").

ubject Identification Code

unique identifier assigned by the investigator to each trial subject to protect e subject's identity and used in lieu of the subject's name when the investigator ports adverse events and/or other trial related data.

rial Site

he location(s) where trial-related activities are actually conducted.

nexpected Adverse Drug Reaction

n adverse reaction, the nature or severity of which is not consistent with the pplicable product information (e.g., Investigator's Brochure for an unapproved ivestigational product or package insert/summary of product characteristics r an approved product) (see the ICH Guideline for Clinical Safety Data lanagement: Definitions and Standards for Expedited Reporting).

ulnerable Subjects

dividuals whose willingness to volunteer in a clinical trial may be unduly fluenced by the expectation, whether justified or not, of benefits associated ith participation, or of a retaliatory response from senior members of a hierarchy case of refusal to participate. Examples are members of a group with a erarchical structure, such as medical, pharmacy, dental, and nursing students, ibordinate hospital and laboratory personnel, employees of the pharmaceutical dustry, members of the armed forces, and persons kept in detention. Other ilnerable subjects include patients with incurable diseases, persons in nursing mes, unemployed or impoverished persons, patients in emergency situations, hnic minority groups, homeless persons, nomads, refugees, minors, and those capable of giving consent.

ell-being (of the trial subjects)

e physical and mental integrity of the subjects participating in a clinical trial.

For Product Safety Concerns and Information please contact our EU
representative GPSR@taylorandfrancis.com
Taylor & Francis Verlag GmbH, Kaufingerstraße 24, 80331 München, Germany

www.ingramcontent.com/pod-product-compliance
Ingram Content Group UK Ltd.
Pitfield, Milton Keynes, MK11 3LW, UK
UKHW020145200926
475702UK00005B/931

* 9 7 8 1 4 6 6 5 9 1 1 7 2 *